Eyewitness
ASTRONOMY

In association with
THE ROYAL OBSERVATORY, GREENWICH

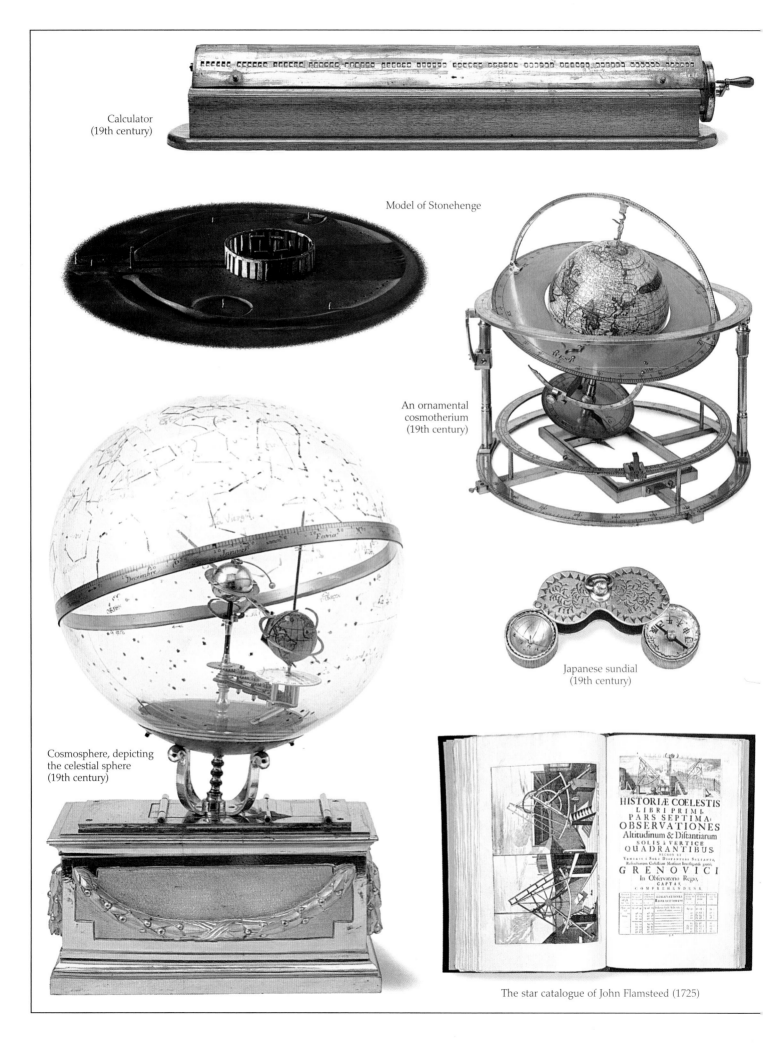

Calculator
(19th century)

Model of Stonehenge

An ornamental
cosmotherium
(19th century)

Cosmosphere, depicting
the celestial sphere
(19th century)

Japanese sundial
(19th century)

The star catalogue of John Flamsteed (1725)

Napier's bones

Eyewitness
ASTRONOMY

Prisms used in a
19th-century spectroscope

Written by
KRISTEN LIPPINCOTT

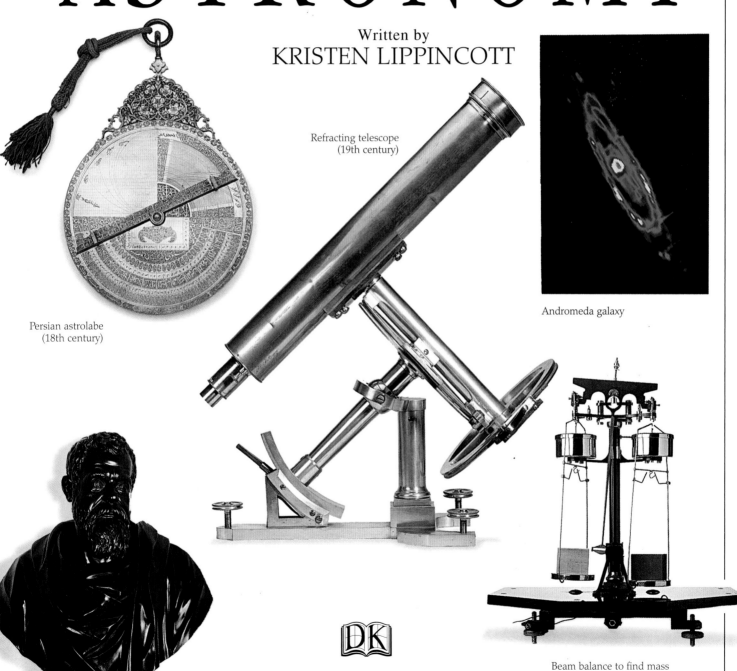

Persian astrolabe
(18th century)

Refracting telescope
(19th century)

Andromeda galaxy

Beam balance to find mass

Bust of Galileo

DK

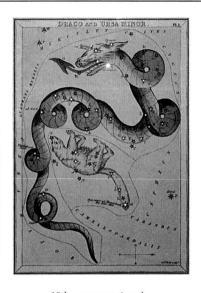

19th-century printed
constellation card

Micrometer for use with
a telescope

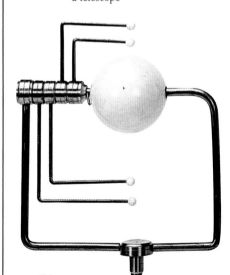

19th-century orrery
showing Uranus with
its four known satellites

LONDON, NEW YORK,
MELBOURNE, MUNICH, and DELHI

Project editor Charyn Jones
Art editor Ron Stobbart
Design assistant Elaine C Monaghan
Production Meryl Silbert
Picture research Becky Halls, Deborah Pownall
Managing editor Josephine Buchanan
Managing art editor Lynne Brown
Special photography Tina Chambers, Clive Streeter
Editorial consultant Dr Heather Couper

THIS EDITION
Editors Clare Hibbert, Sue Nicholson,
Victoria Heywood-Dunne, Marianne Petrou
Art editors Rebecca Johns, David Ball
Senior editor Shaila Awan
Managing editors Linda Esposito, Camilla Hallinan
Managing art editors Jane Thomas, Martin Wilson
Publishing Manager Sunita Gahir
Production editors Siu Yin Ho, Andy Hilliard
Production controllers Jenny Jacoby, Pip Tinsley
Picture research Bridget Tily, Jenny Baskaya, Harriet Mills
DK picture library Rose Horridge, Myriam Megharbi, Emma Shepherd
Consultants Robin Scagell, Dr Jacqueline Mitton

This Eyewitness ® Guide has been conceived by
Dorling Kindersley Limited and Editions Gallimard

First published in Great Britain in 1994.
This revised edition published in 2004, 2008
by Dorling Kindersley Limited,
80 Strand, London WC2R 0RL

A CIP catalogue record for this book is
available from the British Library

ISBN: 978-1-40532-931-6

Colour reproduction by Colourscan, Singapore
Printed and bound by Leo Paper Products Ltd, China

Discover more at
www.dk.com

Compass (19th century)

Drawing an ellipse

A demonstration to
show how different
elements behave in
the Solar System

Contents

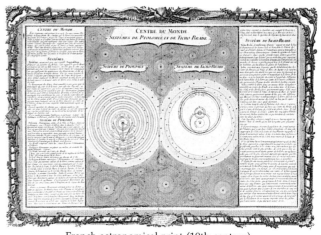

French astronomical print (19th century)

The study of the heavens

THE WORD "ASTRONOMY" comes from a combination of two Greek words: *astron* meaning "star" and *nemein* meaning "to name". Even though the beginnings of astronomy go back thousands of years before the ancient Greeks began studying the stars, the science of astronomy has always been based on the same principle of "naming the stars". Many of the names come directly from the Greeks, since they were the first astronomers to make a systematic catalogue of all the stars they could see. A number of early civilizations remembered the relative positions of the stars by putting together groups that seemed to make patterns in the night sky. One of these looked like a curling river, so it was called Eridanus, the great river; another looked like a hunter with a bright belt and dagger and was called Orion, the hunter (p.61). Stars were named according to their placement inside the pattern and graded according to brightness. For example, the brightest star in the constellation Scorpius is called α Scorpii, because α is the first letter in the Greek alphabet. It is also called Antares, which means "the other Mars", because it shines bright red in the night sky and strongly resembles the blood-red planet, Mars (pp.48–49).

WATCHING THE SKIES
The earliest astronomers were shepherds who watched the heavens for signs of the changing seasons. The clear nights would have given them opportunity to recognize familiar patterns and movements of the brightest heavenly bodies.

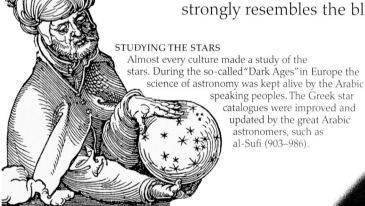

STUDYING THE STARS
Almost every culture made a study of the stars. During the so-called "Dark Ages" in Europe the science of astronomy was kept alive by the Arabic-speaking peoples. The Greek star catalogues were improved and updated by the great Arabic astronomers, such as al-Sufi (903–986).

An engraving of al-Sufi with a celestial globe

UNCHANGING SKY
In all but the largest cities, where the stars are shrouded by pollution or hidden by the glare of street lamps, the recurring display of the night sky is still captivating. The view of the stars from the Earth has changed remarkably little during the past 10,000 years. The sky on any night in the 21st century is nearly the same as the one seen by people who lived thousands of years ago. The night sky for people of the early civilizations would have been more accessible because their lives were not as sheltered from the effects of nature as ours are. Despite the advances in the technology of astronomical observation, which include radio telescopes where the images appear on a computer screen, and telescopes launched into space to detect radiations that do not penetrate our atmosphere, there are still things the amateur astronomer can enjoy. Books and newspapers print star charts so that on a given night, in a specified geographical location, anyone looking upwards into a clear sky can see the constellations for themselves.

TRADITIONAL SYMBOLS

The heritage of the Greek science of the stars passed through many different civilizations. In each case, the figures of the constellations took on the personalities of the heroes of local legends. The Mediterranean animals of the zodiac were transformed by other cultures, such as the Persians and Indians, into more familiar creatures, like the ibex, Brahma bulls, or a crayfish. This page is from an 18th-century Arabic manuscript. It depicts the zodiacal signs of Gemini, Cancer, Aries, and Taurus. The signs are in the Arabic script which is read from right to left.

Rays of light enter the objective lens

Two prisms fold up the light path

LOOKING AT STARS

Many of the sky's mysteries can be seen with a good pair of binoculars. This modern pair gives a better view of the heavens than either Newton, Galileo, or other great astronomers could have seen with their best telescopes (pp.20–21).

Light passes to the eye

FROM SUPERSTITION TO SCIENCE

The science of astronomy grew out of a belief in astrology (pp.16–17), the power of the planets and stars to affect life on the Earth. Each planet was believed to have the personality and powers of one of the gods. Mars, the god of war, shown here, determined war, plague, famine, and violent death.

Quetzalcoatl

AZTEC MYTHOLOGY

In the Americas, the mythology of the stars was stronger than it was in Europe and Asia. This Aztec calendar shows the god Quetzalcoatl, who combined the influences of the Sun and Venus. His worship included ritual human sacrifice.

IMAGING SPACE

With large telescopes, such as the Hubble Space Telescope (HST), astronomers today can observe objects a billion times fainter than anything the ancients saw with the naked eye, including galaxies billions of light years (p.60) away. The HST was put into Earth orbit by the Space Shuttle in 1990. Working above the atmosphere, it can make high-resolution observations in the infrared and ultraviolet as well as of visible light. Astronauts have repaired it several times. If repairs planned for 2008 are successful, HST should keep operating until about 2013.

Ancient astronomy

By watching the cyclic motion of the Sun, the Moon, and the stars, early observers soon realized that these repeating motions could be used to fashion the sky into a clock (to tell the passage of the hours of the day or night) and a calendar (to mark the progression of the seasons). Ancient monuments, such as Stonehenge in the UK and the pyramids of the Maya in Central America, offer evidence that the basic components of observational astronomy have been known for at least 6,000 years. With few exceptions, all civilizations have believed that the steady movements of the sky were the signal of some greater plan. The phenomenon of a solar eclipse (pp.38–39), for example, was believed by some ancient civilizations to be a dragon eating the Sun. A great noise would successfully frighten the dragon away.

DEFYING THE HEAVENS
The ancient poets warn that you should never venture out to sea until the constellation of the Pleiades rises with the Sun in early May. If superpower leaders Mikhail Gorbachov and George Bush Senior had remembered their Greek poets, they would have known better than to try to meet on a boat in the Mediterranean in December 1989. Their summit was almost cancelled because of bad weather.

PHASES OF THE MOON
The changing face of the Moon has always deeply affected people. The New Moon was considered the best time to start an enterprise and the Full Moon was often feared as a time when spirits were free to roam. The word "lunatic" comes from the Latin name for the Moon, *luna*, because it was believed that the rays of the Full Moon caused insanity.

NAMING THE PLANETS
The spread of knowledge tends to follow the two routes of trade and war. As great empires expanded, they brought their gods, customs, and learning with them. The earliest civilizations believed that the stars and planets were ruled by the gods. The Babylonians, for example, named each planet after the god that had most in common with that planet's characteristics. The Greeks and the Romans adopted the Babylonian system, replacing the names with those of their own gods. All the planet names can be traced directly to the Babylonian planet-gods: Nergal has become Mars, and Marduk has become the god Jupiter.

The Roman god Jupiter

Station stone

Aubrey holes are round pits that were part of the earliest structure

THE WORLD'S OLDEST OBSERVATORY
The earliest observatory to have survived is the Chomsung dae Observatory in Kyongju, Korea. A simple, beehive structure, with a central opening in the roof, it resembles a number of prehistoric structures found all over the world. Many modern observatories (pp.26–27) still have a similar roof opening.

RECORDING THE SUN'S MOVEMENTS
Even though the precise significance of the standing stones at Stonehenge remains the subject of debate, it is clear from the arrangement of the stones that it was erected by prehistoric peoples specifically to record certain key celestial events, such as the summer and winter solstices and the spring and autumnal equinoxes. Although Stonehenge is the best known of the ancient megalithic monuments (those made of stone in prehistoric times), the sheer number of similar sites throughout the world underlines how many prehistoric peoples placed an enormous importance on recording the motions of the Sun and Moon.

Back of a Persian astrolabe, 1707

Degree scale

Sight hole

Rotating alidade

Shadow square

BABYLONIAN RECORDS
The earliest astronomical records are in the form of clay tablets from ancient Mesopotamia and the great civilizations that flourished in the plains between the Tigris and Euphrates Rivers for more than 2,000 years. The oldest surviving astronomical calculations are relatively late, dating from the 4th century BCE, but they are clearly based on generations of astronomical observations.

Calendar scale

THE ASTROLABE
One of the problems faced by ancient astronomers was how to simplify the complex calculations needed to predict the positions of the planets and stars. One useful tool was the astrolabe, whose different engraved plates reproduce the sphere of the heavens in two dimensions. The alidade with its sight holes is used to measure the height of the Sun or the stars. By setting this against the calendar scale on the outside of the instrument, a number of different calculations can be made.

PLANNING THE HARVEST
For nearly all ancient cultures the primary importance of astronomy was as a signal of seasonal changes. The Egyptians knew that when the star Sirius rose ahead of the Sun, the annual flooding of the Nile was not far behind. Schedules for planting and harvesting were all set by the Sun, the Moon, and the stars.

Arabic manuscript from the 14th century showing an astrolabe being used

Hele stone marks the original approach to Stonehenge

Avenue

Sun

Slaughter stone formed a ceremonial doorway

Altar stone

Station stone

Barrow

Circular bank and ditch

Circle of Sarsen stones with lintels

Ordering the Universe

A GREAT DEAL OF OUR KNOWLEDGE about the ancient science of astronomy comes from the Alexandrian Greek philosopher Claudius Ptolemaeus (c. 100–178 CE), known as Ptolemy. He was an able scientist in his own right but, most importantly, he collected and clarified the work of all the great astronomers who had lived before him. He left two important sets of books. The *Almagest* was an astronomy textbook that provided an essential catalogue of all the known stars, updating Hipparchus. In the *Tetrabiblos*, Ptolemy discussed astrology. Both sets of books were the undisputed authority on their respective subjects for 1,600 years. Fortunately, they were translated into Arabic because, with the collapse of the Roman Empire around the 4th century, much accumulated knowledge disappeared as libraries were destroyed and books burned.

STAR CATALOGUER
Hipparchus (190–120 BCE) was one of the greatest of the Greek astronomers. He catalogued over 1,000 stars and developed the mathematical science of trigonometry. Here he is looking at the sky through a tube to help him isolate stars – the telescope was not yet invented (pp.22–25).

Julius Caesar

THE LEAP YEAR
One of the problems confronting the astronomer-priests of antiquity was the fact that the lunar year and the solar year (p.13) did not match up. By the middle of the 1st century BCE, the Roman calendar was so mixed up that Julius Caesar (100–44 BCE) ordered the Greek mathematician Sosigenes to develop a new system. He came up with the idea of a leap year every four years. This meant that the odd quarter day of the solar year was rationalized every four years.

Sirius, the dog star

Facsimile (1908) of the Behaim terrestrial globe

Europe
Red Sea

Ocean

Africa

FARNESE ATLAS
Very few images of the constellations have survived from antiquity. The main source for our knowledge is this 2nd-century Roman copy of an earlier Greek statue. The marble statue has the demi-god Atlas holding the heavens on his shoulders. All of the 48 Ptolemaic constellations are clearly marked in low relief.

SPHERICAL EARTH
The concept of a spherical Earth can be traced back to Greece in the 6th century BCE. By Ptolemy's time, astronomers were accustomed to working with earthly (terrestrial) and starry (celestial) globes. The first terrestrial globe to be produced since antiquity, the 15th-century globe by Martin Behaim, shows an image of the Earth that is half-based on myth. The Red Sea, for example, is coloured red.

Navis, the Ship

Atlas

ARABIC SCHOOL OF ASTRONOMY
During the "Dark Ages" the great civilizations of Islam continued to develop the science of astronomy. Ulugh Beigh (c. 15th century) set up his observatory on this site in one of Asia's oldest cities – Samarkand, Uzbekistan. Here, measurements were made with the naked eye.

Geocentric Universe

It is logical to make assumptions from what your senses tell you. From the Earth it looks as if the heavens are circling over our heads. There is no reason to assume the Earth is moving at all. Ancient philosophers, naturally, believed that their Earth was stable and the centre of the great cosmos. The planets were arranged in a series of layers, with the starry heavens – or the fixed stars as they were called – forming a large crystalline casing.

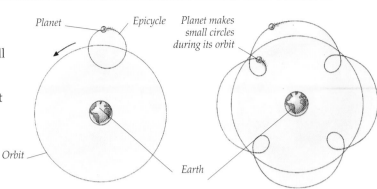

Planet   Epicycle   Planet makes small circles during its orbit

Orbit   Earth

THE EARTH AT THE CENTRE
The geocentric or Earth-centred Universe is often referred to as the Ptolemaic Universe by later scholars to indicate that this was how classical scientists, like the great Ptolemy, believed the Universe was structured. He saw the Earth as the centre of the Universe with the Moon, the known planets, and the Sun moving around it. Aristarchus (c. 310–230 BCE) had already suggested that the Earth travels around the Sun, but his theory was rejected because it did not fit in with the mathematical and philosophical beliefs of the time.

PROBLEMS WITH THE GEOCENTRIC UNIVERSE
The main problem with the model of an Earth-centred Universe was that it did not help to explain the apparently irrational behaviour of some of the planets, which sometimes appear to stand still or move backwards against the background of the stars (p.19). Early civilizations assumed that these odd movements were signals from the gods, but the Greek philosophers spent centuries trying to develop rational explanations for what they saw. The most popular was the notion of epicycles. The planets moved in small circles (epicycles) on their orbits as they circled the Earth.

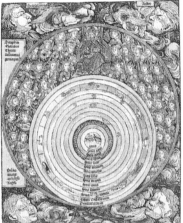

Engraving (1490)
of the Ptolemaic Universe

TEACHING TOOL
Astronomers have always found it difficult to explain the three-dimensional motions of the heavens. Ptolemy used something like this armillary sphere to do his complex astronomical calculations and to pass these ideas on to his students.

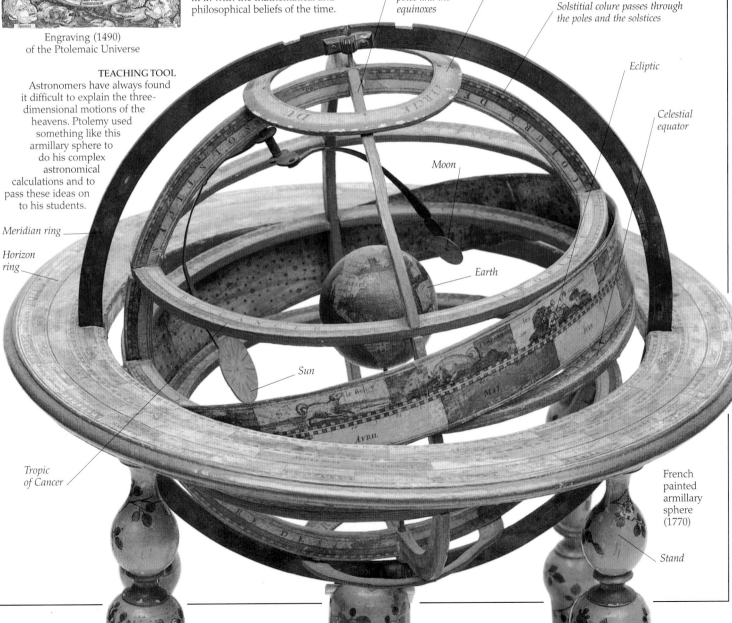

Equinoctial colure passes through the poles and the equinoxes

Arctic circle

Solstitial colure passes through the poles and the solstices

Ecliptic

Celestial equator

Moon

Meridian ring

Horizon ring

Earth

Sun

Tropic of Cancer

French painted armillary sphere (1770)

Stand

The celestial sphere

THE POSITIONS OF ALL OBJECTS IN SPACE are measured according to specific celestial coordinates. The best way to understand the cartography, or mapping, of the sky is to recall how the ancient philosophers imagined the Universe was shaped. They had no real evidence that the Earth moves, so they concluded that it was stationary and that the stars and planets revolve around it. They could see the stars wheeling around a single point in the sky and assumed that this must be one end of the axis of a great celestial sphere. They called it a crystalline sphere, or the sphere of fixed stars, because none of the stars seemed to change their positions relative to each other. The celestial coordinates used today come from this old-fashioned concept of a celestial sphere. The starry (celestial) and earthly (terrestrial) spheres share the same coordinates, such as a North and South Pole and an equator.

STAR TRAILS
A long photographic exposure of the sky taken from the northern hemisphere of the Earth shows the way in which stars appear to go in circles around the Pole Star or Polaris. Polaris is a bright star that lies within 1° of the true celestial pole, which, in turn, is located directly above the North Pole of the Earth. The rotation of the Earth on its north/south axis is the reason why the stars appear to move across the sky. Those closer to the Poles appear to move less than those further away.

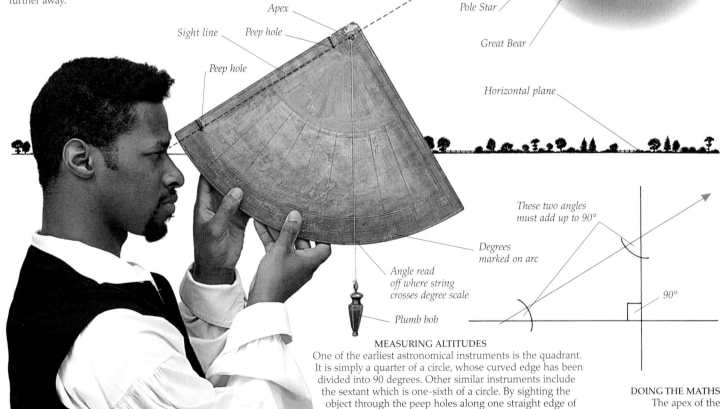

Apex

Sight line

Peep hole

Peep hole

Pole Star

Great Bear

Horizontal plane

These two angles must add up to 90°

Degrees marked on arc

Angle read off where string crosses degree scale

Plumb bob

90°

MEASURING ALTITUDES
One of the earliest astronomical instruments is the quadrant. It is simply a quarter of a circle, whose curved edge has been divided into 90 degrees. Other similar instruments include the sextant which is one-sixth of a circle. By sighting the object through the peep holes along one straight edge of the quadrant, the observer can measure the height, or altitude, of that object. The altitude is the height in degrees (°) of a star above the horizon; it is not a linear measurement. A string with a plumb bob falls from the apex of the quadrant so that it intersects the divided arc. Since the angle between the vertical of the plumb bob and the horizontal plane of the horizon is 90°, simple mathematics can be used to work out the angle of the altitude.

DOING THE MATHS
The apex of the quadrant is a 90° angle. As the sum of the angles of a triangle adds up to 180°, this means that the sum of the other two angles must add up to 90° too.

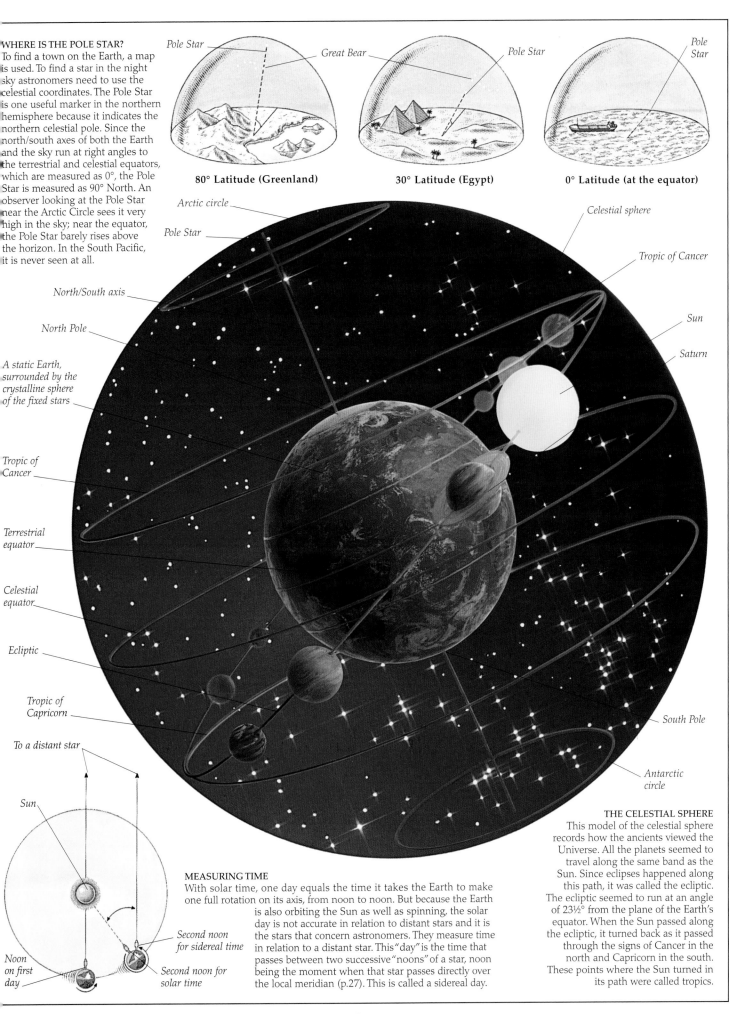

WHERE IS THE POLE STAR?
To find a town on the Earth, a map is used. To find a star in the night sky astronomers need to use the celestial coordinates. The Pole Star is one useful marker in the northern hemisphere because it indicates the northern celestial pole. Since the north/south axes of both the Earth and the sky run at right angles to the terrestrial and celestial equators, which are measured as 0°, the Pole Star is measured as 90° North. An observer looking at the Pole Star near the Arctic Circle sees it very high in the sky; near the equator, the Pole Star barely rises above the horizon. In the South Pacific, it is never seen at all.

Pole Star — Great Bear

80° Latitude (Greenland)

Pole Star

30° Latitude (Egypt)

Pole Star

0° Latitude (at the equator)

Arctic circle

Pole Star

Celestial sphere

Tropic of Cancer

North/South axis

Sun

North Pole

Saturn

A static Earth, surrounded by the crystalline sphere of the fixed stars

Tropic of Cancer

Terrestrial equator

Celestial equator

Ecliptic

Tropic of Capricorn

South Pole

To a distant star

Antarctic circle

Sun

Second noon for sidereal time

Noon on first day

Second noon for solar time

MEASURING TIME
With solar time, one day equals the time it takes the Earth to make one full rotation on its axis, from noon to noon. But because the Earth is also orbiting the Sun as well as spinning, the solar day is not accurate in relation to distant stars and it is the stars that concern astronomers. They measure time in relation to a distant star. This "day" is the time that passes between two successive "noons" of a star, noon being the moment when that star passes directly over the local meridian (p.27). This is called a sidereal day.

THE CELESTIAL SPHERE
This model of the celestial sphere records how the ancients viewed the Universe. All the planets seemed to travel along the same band as the Sun. Since eclipses happened along this path, it was called the ecliptic. The ecliptic seemed to run at an angle of 23½° from the plane of the Earth's equator. When the Sun passed along the ecliptic, it turned back as it passed through the signs of Cancer in the north and Capricorn in the south. These points where the Sun turned in its path were called tropics.

The uses of astronomy

WITH ALL THE TOOLS OF MODERN TECHNOLOGY, it is sometimes hard to imagine how people performed simple functions such as telling the time or knowing where they were on the Earth before the invention of clocks, maps, or navigational satellites. The only tools available were those provided by nature. The astronomical facts of the relatively regular interval of the day, the constancy of the movements of the fixed stars, and the assumption of certain theories, such as a spherical Earth, allowed people to measure their lives. By calculating the height of the Sun or certain stars, the ancient Greeks began to understand the shape and size of the Earth. In this way, they were able to determine their latitude. By plotting coordinates against a globe, they could fix their position on the Earth's surface. And by setting up carefully measured markers, or gnomons, they could begin to calculate the time of day.

Sun
Alexandria
Syene

MEASURING THE EARTH
About 230 BCE Eratosthenes (c. 270–190 BCE) estimated the size of the Earth by using the Sun. He discovered that the Sun was directly above his head at Syene (present-day Aswan) in Upper Egypt at noon on the summer solstice. In Alexandria, directly north, the Sun was about 7° from its highest point (the zenith) at the summer solstice. Since Eratosthenes knew that the Earth was spherical (360° in circumference), the distance between the two towns should be 7/360ths of the Earth's circumference.

Latitude scale
Moveable cursor
Sight hole
Sight hole
Hour scale
Zodiac scale
Plumb bob

AN ANCIENT SUNDIAL
Very early on, people realized that they could keep time by the Sun. Simple sundials like this allowed the traveller or merchant to know the local time for several different towns during a journey. The altitude of the Sun was measured through the sight holes in the bow and stern of the "little ship". When the cursor on the ship's mast was set to the correct latitude, the plumb bob would fall on the proper time.

HOW A SUNDIAL WORKS
As the Sun travels across the sky, the shadow it casts changes in direction and length. A sundial works by setting a gnomon, or "indicator", so that the shadow the Sun casts at noon falls due north/south along a meridian. (A meridian is an imaginary line running from pole to pole; another name for meridian is a line of longitude.) The hours can then be divided before and after the noon mark. The terms "a.m." and "p.m." for morning and afternoon come from the Latin words meaning before and after the Sun passes the north/south meridian (*ante meridiem* and *post meridiem*).

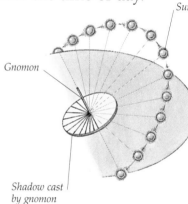

Sun
Gnomon
Shadow cast by gnomon at noon

FINDING MECCA
Part of Islamic worship is regular prayers, in which the faithful face toward the holy city of Mecca. The qiblah (direction of Mecca) indicator is a sophisticated instrument, developed during the Middle Ages to find the direction of Mecca. It also uses the Sun to determine the time for beginning and ending prayers.

Towns with their latitudes
Qiblah lid

CROSSING THE SOUTH PACIFIC
It was thought that the early indigenous peoples of Polynesia were too "primitive" to have sailed the great distances between the north Pacific Ocean and New Zealand in the south. However, many tribes, including the Maoris, were capable of navigating thousands of kilometres using only the stars to guide them.

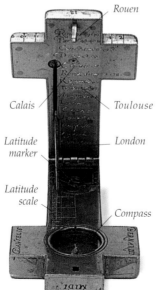

Rouen
Calais
Toulouse
Latitude marker
London
Latitude scale
Compass

CRUCIFORM SUNDIAL
Travelling Christian pilgrims often worried that any ornament might be considered a symbol of vanity. They solved this problem by incorporating religious symbolism into their sundials. This dial, shaped in the form of a cross, provided the means for telling the time in a number of English and French towns.

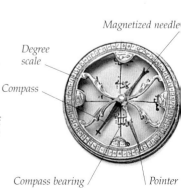

Magnetized needle
Degree scale
Compass
Compass bearing
Pointer

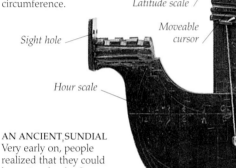

Argo, the ship

A CELESTIAL GLOBE
The celestial globe records the figures and stars of all the constellations against a grid of lines representing longitude and latitude. During the 17th and 18th centuries all ships of the Dutch East India Company were given a matching pair of globes – terrestrial (p.10) and celestial. Calculations could be made by comparing the coordinates on the two different globes. In practice, however, most navigators seemed to use flat sea-charts to plot their journeys.

Hydra, the water snake

Meridian ring

Celestial globe 1618

The Southern Cross

Southern triangle

Centaurus, the Centaur

Sun

THE GREAT NAVIGATORS
Explorers of the 16th century had no idea what they would find when they set out to sea. Their heads were full of fables about mermaids and sea monsters. Even though this engraving of the Portuguese navigator Ferdinand Magellan (1480–1521) has many features that are clearly fantastical, it does show him using a pair of dividers to measure off an armillary sphere (p.11). Beside the ship, the Sun God Apollo shines brightly; it was usually the Sun's position in the sky that helped a navigator find his latitude.

Shadow vane lined up with horizon vane

Horizon vane

Scale in degrees

Holder

Sight vane

Navigator with his back to the Sun

Horizon

Scale in degrees

USING A BACKSTAFF
The backstaff allowed a navigator to measure the height of the Sun without having to stare directly at it. The navigator held the instrument so that the shadow cast by the shadow vane would fall directly on to the horizon vane. Moving the sight vane, the navigator lined it up so he could see the horizon through the sight vane and the horizon vane. By adding together the angles of the sight and shadow vanes, the navigator could calculate the altitude of the Sun, from which he could determine the precise latitude of his ship.

Two angles give the Sun's altitude

90° angle

Horizon

DOING THE MATHEMATICS
To work out latitude at sea, a navigator needs to find the altitude of the Sun at noon. He doesn't even need to know the time; as long as the Sun is at its highest point in the sky, the altitude can be measured with a backstaff or other instrument (p.12). Then, using nautical tables of celestial coordinates, he can find his latitude with a simple equation using the angle of altitude and the coordinates of the Sun in the celestial sphere (p.13).

Astrology

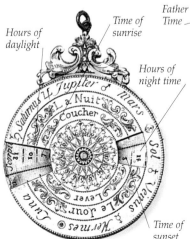

THE ASTROLOGER
In antiquity, the astrologers' main task was to predict the future. This woodcut, dating from 1490, shows two astrologers working with arrangements of the Sun, Moon, and planets to find the astrological effects on people's lives.

THE WORD "ASTROLOGY" comes from the Greek *astron* meaning "star" and *logos* meaning "the science". Since Babylonian times, people staring at the night sky were convinced that the regular motions of the heavens were indications of some great cosmic purpose. Priests and philosophers believed that if they could map the stars and the movements of the stars, they could decode these messages and understand the patterns that had an effect on past and future events. What was originally observational astronomy – observing the stars and planets – gradually grew into the astrology that has today become a regular part of many people's lives. However, there is no evidence that the stars and planets have any effect on our personalities or our destinies. Astronomers now think astrology is superstition. Its original noble motives should not be forgotten, however. For most of the so-called "Dark Ages", when all pure science was in deep hibernation, it was astrology and the desire to know about the future that kept the science of astronomy alive.

RULERSHIP OVER ORGANS
Until the discoveries of modern medicine, people believed that the body was governed by four different types of essences called "humours". An imbalance in these humours would lead to illness. Each of the 12 signs of the zodiac (above) had special links with each of the humours and with parts of the human body. So, for example, for a headache due to moisture in the head (a cold), treatment would be with a drying agent – some plant ruled by the Sun or an "Earth-sign", like Virgo – when a New Moon was well placed towards the sign of Aries, which ruled the head.

Dates in the month

Days in the week

Father Time

Back of calendar

Hours of daylight

Time of sunrise

Father Time

Hours of night time

Time of sunset

PERPETUAL CALENDAR
The names for the days of the week show traces of astrological belief – for example, Sunday is the Sun's day, and Monday is the Moon's day. This simple perpetual calendar, which has small planetary signs next to each day, shows the day of the week for any given date. The user can find the day by turning the inner dial to a given month or date and reading off the information.

LEO, THE LION
These 19th-century French constellation cards show each individual star marked with a hole through which light shines. Astrologically, each zodiacal sign has its own properties and its own friendships and enemies within the zodiacal circle. Each sign is also ruled by a planet, which similarly has its own properties, friendships, and enemies. So, for example, a person born while the Sun is passing through Leo is supposed to be kingly, like a lion.

PLANETARY POSITIONS

One way in which planets are supposed to be in or out of harmony with one another depends on their relative positions in the heavens. When two planets are found within a few degrees of each other, they are said to be in conjunction. When planets are separated by exactly 180° in the zodiacal band, they are said to be in opposition.

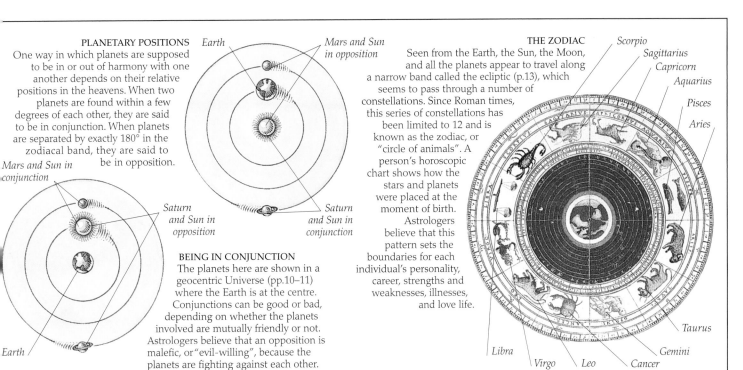

Earth

Mars and Sun in opposition

Mars and Sun in conjunction

Saturn and Sun in opposition

Saturn and Sun in conjunction

Earth

BEING IN CONJUNCTION

The planets here are shown in a geocentric Universe (pp.10–11) where the Earth is at the centre. Conjunctions can be good or bad, depending on whether the planets involved are mutually friendly or not. Astrologers believe that an opposition is malefic, or "evil-willing", because the planets are fighting against each other.

THE ZODIAC

Seen from the Earth, the Sun, the Moon, and all the planets appear to travel along a narrow band called the ecliptic (p.13), which seems to pass through a number of constellations. Since Roman times, this series of constellations has been limited to 12 and is known as the zodiac, or "circle of animals". A person's horoscopic chart shows how the stars and planets were placed at the moment of birth. Astrologers believe that this pattern sets the boundaries for each individual's personality, career, strengths and weaknesses, illnesses, and love life.

Scorpio
Sagittarius
Capricorn
Aquarius
Pisces
Aries
Taurus
Gemini
Cancer
Leo
Virgo
Libra

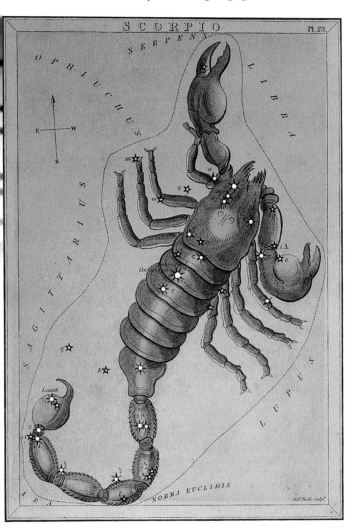

SCORPIO, THE SCORPION

Most of the constellations are now known by the Latinized versions of their original Greek names. This card shows Scorpius, or Scorpio. This is the sign "through which" the Sun passes between late October and late November. Astrologers believe that people born during this time of year are intuitive, yet secretive, like a scorpion scuttling under a rock.

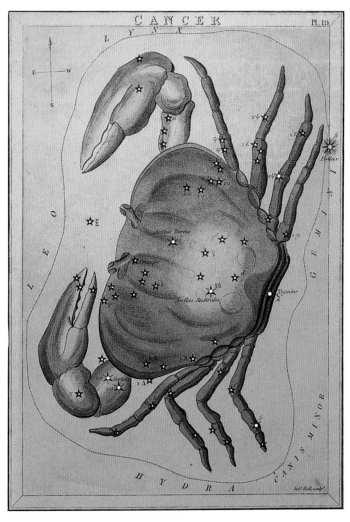

CANCER, THE CRAB

Someone who is born while the Sun is transitting the constellation of Cancer is supposed to be a home-body, like a crab in its shell. These hand-painted cards are collectively known as *Urania's Mirror* – Urania is the name of the muse of astronomy (p.19). By holding the cards up to the light, it is possible to learn the shapes and relative brightnesses of the stars in each constellation.

The Copernican revolution

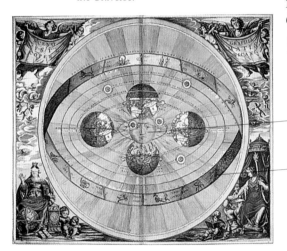

NICOLAUS COPERNICUS
The Polish astronomer Nicolaus Copernicus (1473–1543) made few observations. Instead, he read the ancient philosophers and discovered that none of them had been able to agree about the structure of the Universe.

IN 1543 NICOLAUS COPERNICUS published a book that changed the perception of the Universe. In his *De revolutionibus orbium coelestium* ("Concerning the revolutions of the celestial orbs"), Copernicus argued that the Sun, and not the Earth, is at the centre of the Universe. It was a heliocentric Universe, *helios* being the Greek word for Sun. His reasoning was based on the logic of the time. He argued that a sphere moves in a circle that has no beginning and no end. Since the Universe and all the heavenly bodies are spherical, their motions must be circular and uniform. In the Ptolemaic, Earth-centred system (pp.10–11), the paths of the planets are irregular. Copernicus assumed that uniform motions in the orbits of the planets appear irregular to us because the Earth is not at the centre of the Universe. These discoveries were put forward by many different astronomers, but they ran against the teachings of both the Protestant and Catholic churches. In 1616 all books written by Copernicus and any others that put the Sun at the centre of the Universe were condemned by the Catholic Church.

COPERNICAN UNIVERSE
Copernicus based the order of his Solar System on how long it took each planet to complete a full orbit. This early print shows the Earth in orbit around the Sun with the zodiac beyond.

Sun

Zodiac

THE GREAT OBSERVER
In 1672, the Danish astronomer Tycho Brahe (1546–1601) discovered a bright new star in the constellation Cassiopeia. It was what astronomers today call a "supernova" (p.61). It was so bright that it was visible even during the day. This appearance challenged the inherited wisdom from the ancients, which claimed that the stars were eternal and unchanging. To study what this appearance might mean, a new observatory was set up near Copenhagen in Denmark. Brahe remeasured 788 stars of Ptolemy's great star catalogue, thereby producing the first complete, modern stellar atlas.

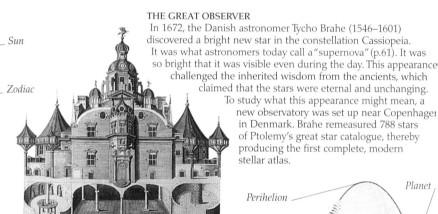

Uranibourg, Tycho's observatory on the island of Hven

DRAWING AN ELLIPSE
An ellipse can be drawn by pushing two pins into a board and linking them with a loop of thread. When a pencil is placed within the loop and moved around the pins, keeping the loop tight, the shape it makes is an ellipse. The position of each pin is called a focus. In the Solar System the Sun is at one focus of the ellipse in a planetary orbit. The wider apart the pins are placed, the more eccentric the planetary orbit (pp.36–37).

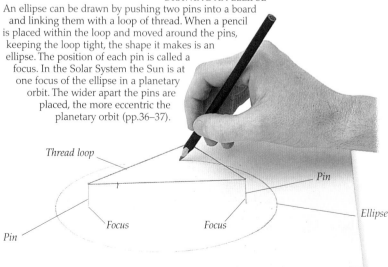

Thread loop

Pin

Ellipse

Pin

Focus

Focus

LAWS OF PLANETARY MOTION
Johannes Kepler (above right) added the results of his own observations to Tycho's improved planetary and stellar measurements. Kepler discovered that the orbits of the planets were not perfectly circular, as had been believed for 1,600 years. They were elliptical, with the Sun placed at one focus of the ellipse (left). While observing the orbit of Mars, Kepler discovered that there are variations in its speed. At certain points in its orbit, Mars seemed to be travelling faster than at other times. He soon realized that the Sun was regulating the orbiting speed of the planet. When it is closest to the Sun - its perihelion – the planet orbits most quickly; at its aphelion - furthest from the Sun – it slows down.

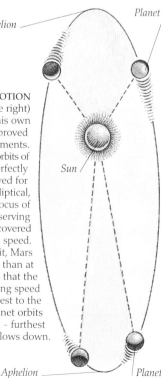

Planet

Perihelion

Sun

Aphelion

Planet

JOHANNES KEPLER (1571–1630)
It was due to the intervention of Tycho Brahe that the German mathematician Johannes Kepler landed the prestigious position of Imperial Mathematician in 1601. Tycho left all his papers to Kepler, who was a vigorous supporter of the Copernican heliocentric system. Kepler formulated three laws of planetary motion and urged Galileo (p.20) to publish his research in order to help prove the Copernican thesis.

Planet paths shown in a planetarium

APPARENT PATHS
The irregular motion that disproved the geocentric universe was the retrograde motion of the planets. From an earthly perspective, some of the planets – particularly Mars – seem to double back on their orbits, making great loops in the night sky. (The light display above draws the apparent orbit of Mars.) Ptolemy proposed that retrograde motion could be explained by planets travelling on smaller orbits (p.11). Once astronomers realized that the Sun is the centre of the Solar System, the apparent path of Mars, for example, could be explained. But first it had to be understood that the Earth had a greater orbiting speed than that of Mars, which appeared to slip behind. Even though the orbit of Mars seems to keep pace with the Earth (below left), the apparent path is very different (above left).

Apparent path of Mars   *Line of sight*

Orbit of Mars   *Sun*   A model showing the true and apparent orbits of Mars from an earthly perspective   *Orbit of the Earth*

WEIGHING UP THE THEORIES
This engraving from a 17th-century manuscript shows Urania, the muse of astronomy, comparing the different theoretical systems for the arrangement of the Universe. Ptolemy's system is at her feet, and Kepler's is outweighed by Tycho's system on the right.

Intellectual giants

Iᴛ ᴛᴀᴋᴇs ʙᴏᴛʜ ʟᴜᴄᴋ ᴀɴᴅ ᴄᴏᴜʀᴀɢᴇ to be a radical thinker. Galileo Galilei (1564–1642) had the misfortune of being brilliant at a time when new ideas were considered dangerous. His numerous discoveries, made with the help of the newly invented telescope, provided ample support for the Copernican heliocentric, or Sun-centred, Universe (pp.18–19). Galileo's findings about the satellites of Jupiter (p.50) and the phases of Venus clearly showed that the Earth could not be the centre of all movement in the Universe and that the heavenly bodies were not perfect in their behaviour. For this Galileo was branded a heretic and sentenced to a form of life imprisonment. The great English physicist Isaac Newton (1642–1727), born the year Galileo died, had both luck and courage. He lived in an age enthusiastic for new ideas, especially those related to scientific discovery.

POPE URBAN VIII
Originally, the Catholic Church had welcomed Copernicus's work (pp.18–19). However, by 1563 the Church was becoming increasingly strict and abandoned its previously lax attitude toward any deviation from established doctrine. Pope Urban VIII was one of the many caught in this swing. As a cardinal, he had been friendly with Galileo and often had Galileo's book, *Il Saggiatore*, read to him aloud at meals. In 1635, however, he authorized the Grand Inquisition to investigate Galileo.

GALILEO'S TELESCOPE
Galileo never claimed to have invented the telescope. In *Il Saggiatore*, "The Archer", he commends the "simple spectacle-maker" who "found the instrument" by chance. When he heard of Lippershey's results (p.22), Galileo reinvented the instrument from the description of its effects. His first telescope magnified at eight times. Within a few days, however, he had constructed a telescope with x 20 magnification. He went on to increase his magnification to x 30, having ground the lenses himself.

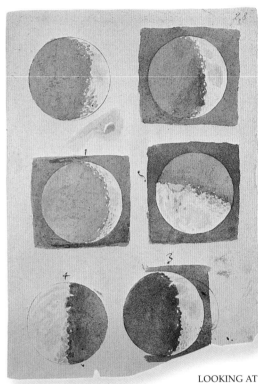

LOOKING AT THE MOON'S SURFACE
Through his telescope, Galileo measured the shadows on the Moon to show how the mountains there were much taller than those on Earth. These ink sketches were published in his book *Sidereus nuncius*, "Messenger of the Stars", in 1610.

RENAISSANCE MAN
In 1611 Galileo travelled to Rome to discuss his findings about the Sun and its position in the Universe with the leaders of the Church. They accepted his discoveries, but not the theory that underpinned them – the Copernican, heliocentric Universe (pp.18–19). Galileo was accused of heresy and, in 1635, condemned for disobedience and sentenced to house arrest until his death in 1642. He was pardoned in 1992.

PHASES OF VENUS
From his childhood days, Galileo was characterized as the sort of person who was unwilling to accept facts without evidence. In 1610, by applying the telescope to astronomy, he discovered the moons of Jupiter and the phases of Venus. He immediately understood that the phases of Venus are caused by the Sun shining on a planet that revolves around it. He knew that this was proof that the Earth was not the centre of the Universe. He hid his findings in a Latin anagram, or word puzzle, as he did with many of the discoveries that he knew would be considered "dangerous" by the authorities.

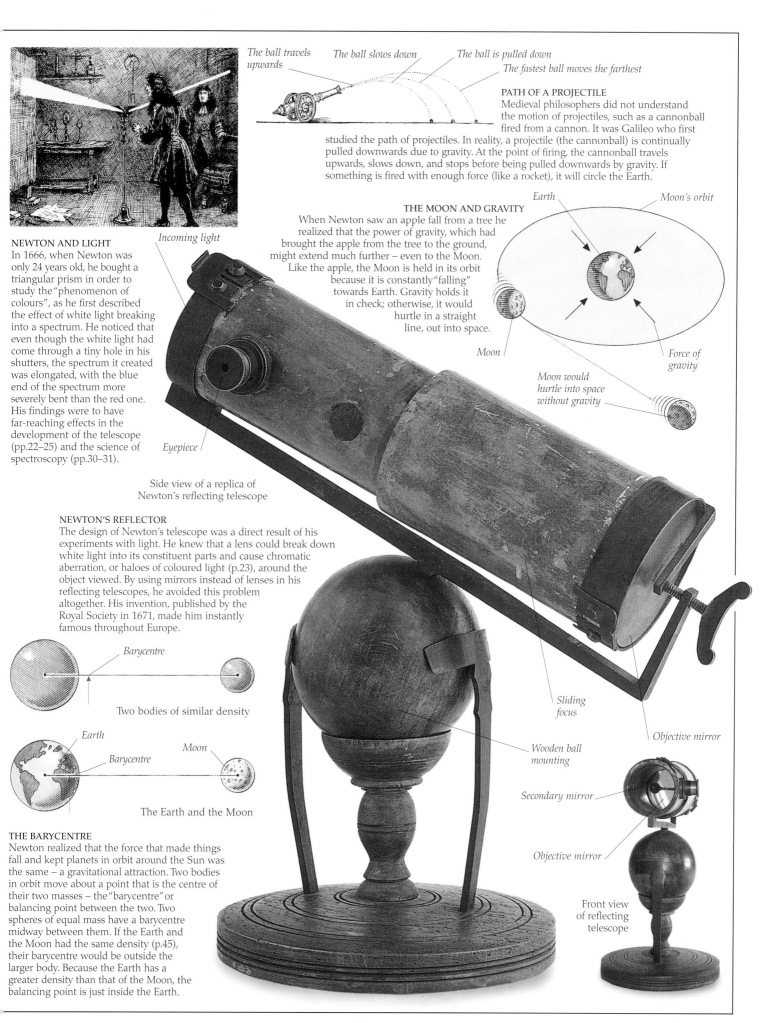

The ball travels upwards   **The ball slows down**   **The ball is pulled down**
The fastest ball moves the farthest

PATH OF A PROJECTILE
Medieval philosophers did not understand the motion of projectiles, such as a cannonball fired from a cannon. It was Galileo who first studied the path of projectiles. In reality, a projectile (the cannonball) is continually pulled downwards due to gravity. At the point of firing, the cannonball travels upwards, slows down, and stops before being pulled downwards by gravity. If something is fired with enough force (like a rocket), it will circle the Earth.

NEWTON AND LIGHT
In 1666, when Newton was only 24 years old, he bought a triangular prism in order to study the "phenomenon of colours", as he first described the effect of white light breaking into a spectrum. He noticed that even though the white light had come through a tiny hole in his shutters, the spectrum it created was elongated, with the blue end of the spectrum more severely bent than the red one. His findings were to have far-reaching effects in the development of the telescope (pp.22–25) and the science of spectroscopy (pp.30–31).

Incoming light

THE MOON AND GRAVITY
When Newton saw an apple fall from a tree he realized that the power of gravity, which had brought the apple from the tree to the ground, might extend much further – even to the Moon. Like the apple, the Moon is held in its orbit because it is constantly "falling" towards Earth. Gravity holds it in check; otherwise, it would hurtle in a straight line, out into space.

Earth   *Moon's orbit*

Moon   *Force of gravity*

Moon would hurtle into space without gravity

Eyepiece

Side view of a replica of Newton's reflecting telescope

NEWTON'S REFLECTOR
The design of Newton's telescope was a direct result of his experiments with light. He knew that a lens could break down white light into its constituent parts and cause chromatic aberration, or haloes of coloured light (p.23), around the object viewed. By using mirrors instead of lenses in his reflecting telescopes, he avoided this problem altogether. His invention, published by the Royal Society in 1671, made him instantly famous throughout Europe.

Barycentre

Two bodies of similar density

Earth   *Moon*
Barycentre

The Earth and the Moon

THE BARYCENTRE
Newton realized that the force that made things fall and kept planets in orbit around the Sun was the same – a gravitational attraction. Two bodies in orbit move about a point that is the centre of their two masses – the "barycentre" or balancing point between the two. Two spheres of equal mass have a barycentre midway between them. If the Earth and the Moon had the same density (p.45), their barycentre would be outside the larger body. Because the Earth has a greater density than that of the Moon, the balancing point is just inside the Earth.

Sliding focus

Wooden ball mounting

Objective mirror

Secondary mirror

Objective mirror

Front view of reflecting telescope

Optical principles

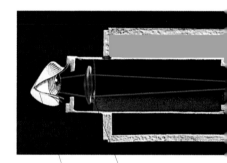

P<small>EOPLE HAVE BEEN AWARE</small> of the magnifying properties of a curved piece of glass since at least 2,000 BCE. The Greek philosopher Aristophanes in the 5th century BCE had used a glass globe filled with water in order to magnify the fine print in his manuscripts. In the middle of the 13th century the English scientist Roger Bacon (1214–1292) proposed that the "lesser segment of a sphere of glass or crystal" will make small objects appear clearer and larger. For this suggestion, Bacon was branded by his colleagues a dangerous magician and imprisoned for ten years. Even though spectacles were invented in Italy some time between 1285 and 1300, superstitions were not overcome for another 250 years when scientists discovered the combination of lenses that would lead to the invention of the telescope. There are two types of telescope. The refracting telescope uses lenses to bend light; the reflecting telescope uses mirrors to reflect the light back to the observer.

Viewer *Convex eyepiece lens*

INVENTOR OF THE TELESCOPE
It is believed that the first real telescope was invented in 1608 in Holland by the spectacle maker Hans Lippershey from Zeeland. According to the story, two of Lippershey's children were playing in his shop and noticed that by holding two lenses in a straight line they could magnify the weather vane on the local church. Lippershey placed the two lenses in a tube and claimed the invention of the telescope. In the mid-1550s an Englishman Leonard Digges had created a primitive instrument that, with a combination of mirrors and lenses, could reflect and enlarge objects viewed through it. There was controversy about whether this was a true scientific telescope or not. It was Galileo (p.20) who adapted the telescope to astronomy.

Decorative ribbons might be attached here

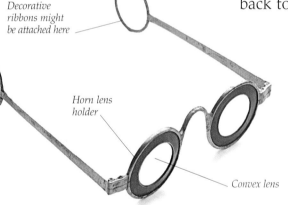

Horn lens holder

Convex lens

EARLY SPECTACLES (1750)
Most early spectacles like these had convex lenses. These helped people who were long-sighted to focus on objects close at hand. Later, spectacles were made with concave lenses for those who were short-sighted.

Water

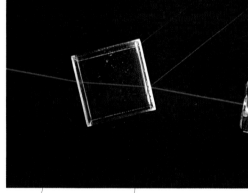

Light from laser

HOW REFRACTION WORKS
Light usually travels in a straight line, but it can be bent or "refracted" by passing it through substances of differing densities. This laser beam (here viewed from overhead) seems to bend as it is directed at a rectangular-shaped container of water because the light is passing through three different media – water, glass, and air.

Path of light is bent again on re-entering air   *Light is bent*

Reflected light beam   *Light from laser is bent back by a shiny surface*   *Incident light beam*

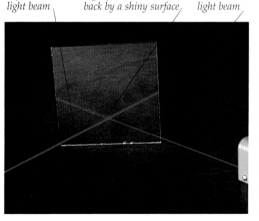

HOW REFLECTION WORKS
The word reflection comes from the Latin *reflectere*, meaning to "bend back". A shiny surface will bend back rays of light that strike it. The rays approaching the mirror are called incident rays and those leaving it are called outgoing, or reflected, rays. The angle at which the incident rays hit the mirror is the same as the angle of the reflected rays leaving it. What the eye sees are the light rays reflected in the mirror.

Large concave mirror

JOHN DOLLOND
The English optician John Dollond (1706–1761) was the first to perfect the achromatic lens so that it might be manufactured more easily and solve the problem of chromatic aberration. Dollond claimed to have invented a new method of refraction.

CHROMATIC ABERRATION
When light goes through an ordinary lens, each colour in the spectrum is bent at a different angle causing rainbows to appear around the images viewed. The blue end of the spectrum will bend more sharply than the red end of the spectrum so that the two colours will focus at different points. This is chromatic aberration. By adding a second lens made from a different kind of glass (and with a different density), all the colours focus at the same point and the problem is corrected.

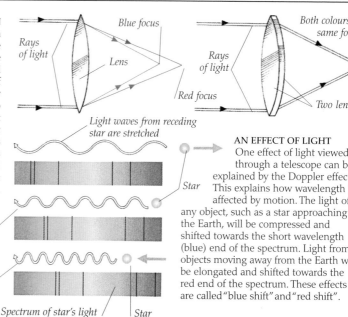

Blue focus

Rays of light

Lens

Red focus

Both colours at same focus

Rays of light

Two lenses

Light waves from receding star are stretched

Earth

Light waves from a stationary star

Light waves from star approaching the Earth are compressed

Star

Spectrum of star's light

Star

AN EFFECT OF LIGHT
One effect of light viewed through a telescope can be explained by the Doppler effect. This explains how wavelength is affected by motion. The light of any object, such as a star approaching the Earth, will be compressed and shifted towards the short wavelength (blue) end of the spectrum. Light from objects moving away from the Earth will be elongated and shifted towards the red end of the spectrum. These effects are called "blue shift" and "red shift".

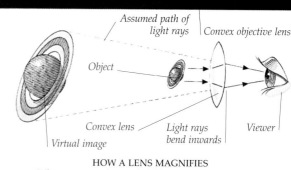

A REFRACTING TELESCOPE
In a refracting telescope the convex objective lens (the one furthest from the eye) collects the light and forms an image. The convex eyepiece lens (the one nearest the eye) magnifies the image in just the same way as a magnifying glass. Galileo used a similar type of refracting telescope (p.20). The main problem with the refracting telescope is chromatic aberration (above).

Viewer

Eyepiece lens

A REFLECTING TELESCOPE
Sir Isaac Newton (p.21) developed a version of the reflecting telescope that consists of a large concave, or curved, mirror to catch the light. The mirror then sends the light back to an inclined flat, or plane, mirror where the image is formed. The eyepiece lens magnifies the image. Unlike the lenses in a refracting telescope, the mirrors in a reflecting telescope do not cause chromatic aberration, so the image is clearer.

Assumed path of light rays

Convex objective lens

Object

Convex lens

Light rays bend inwards

Viewer

Virtual image

HOW A LENS MAGNIFIES
When a convex lens is held between the eye and an object, the object appears larger because the lens bends the rays of light inwards. The eye naturally traces the rays of light back towards the object in straight lines. It sees a "virtual" image, which is larger than the original image. The degree of magnification depends on the angles formed by the curvature of the lens.

Plane mirror

Incoming light

The optical telescope

THE MORE LIGHT THAT REACHES THE EYEPIECE in a telescope, the brighter the image of the heavens will be. Astronomers made their lenses and mirrors bigger, they changed the focal length of the telescopes, and combined honeycombs of smaller mirrors to make a single, large reflective surface in order to capture the greatest amount of light and focus it on to a single point. During the 19th century refracting telescopes (pp.22–23) were preferred and opticians devoted themselves to perfecting large lenses free of blemishes. In the 20th century there were advances in materials and mirror coatings. Large mirrors collect more light than small ones, but are also heavier. They may even sag under their own weight, distorting the image. One solution is segmented-mirror telescopes, where many smaller mirrors are mounted side by side. Another is "active optics", where mirrors move to compensate for any sagging.

CAMERAS ON TELESCOPES
Since the 19th century astronomical photography has been an important tool for astronomers. By attaching a camera to a telescope that has been specially adapted with a motor that can be set to keep the telescope turning at the same speed as the rotation of the Earth, the astronomer can take very long exposures of distant stars (p.12). Before the invention of photography, astronomers had to draw everything they saw. They had to be artists as well as scientists.

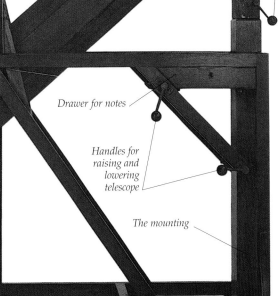

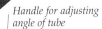

Eyepiece

Eyepiece mounting

Handle for adjusting angle of tube

Guide rails for raising telescope

Telescope tube

Main mirror located inside the tube

Drawer for notes

Handles for raising and lowering telescope

The mounting

Wheeled base

HERSCHEL'S TELESCOPE
First out of economic necessity and later as an indication of his perfectionism, the English astronomer William Herschel (1738–1822) always built his telescopes and hand-ground his own lenses and mirrors. The magnification of a telescope like his 15–cm (6-in) Newtonian reflector is about 200 times. This wooden telescope is the kind he would have used during his great survey of the sky, during which he discovered the planet Uranus (pp.54–55).

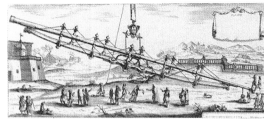

MORE MAGNIFICATION
Increasing the magnification of telescopes was one of the major challenges facing early astronomers. Since the technology to make large lenses was not sufficiently developed, the only answer was to make telescopes with a very long distance between the eyepiece lens and the objective lens. In some instances, this led to telescopes of ridiculous proportions, as shown in this 18th-century engraving. These long focal-length telescopes were impossible to use. The slightest vibration caused by someone walking past would make the telescope tremble so violently that observations were impossible.

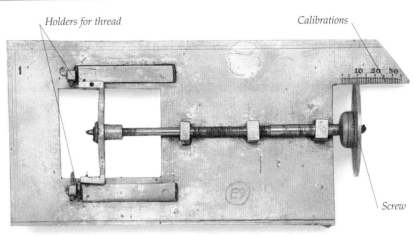

Holders for thread

Calibrations

Screw

MEASURING ACROSS VAST DISTANCES

The bigger the telescope, the larger its scale will be. This means that measurements become increasingly crude. A micrometer can be set to provide extremely fine gradations, a necessary element when measuring the distances between two stars in the sky that are a very long way away. This micrometer was made by William Herschel. To pin-point the location of a star, a fine hair or spider's web was threaded between two holders that were adjusted by means of the finely turned screw on the side.

AN EQUATORIAL MOUNT

Telescopes have to be mounted in some way. The equatorial mount used to be the favoured mount, and is still preferred by amateur astronomers. The telescope is lined up with the Earth's axis, using the Pole Star as a guide. In the southern hemisphere, other stars near the sky's south pole are used. The telescope can swing around this axis, automatically following the tracks of stars in the sky as they circle around the Pole Star. The equatorial mount was used for this 71 cm (28 in) refractor, installed at Greenwich, England in 1893.

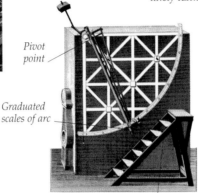

Pivot point

Graduated scales of arc

Ladder for an astronomer to reach the eyepiece

ASTRONOMICAL QUADRANT

Most early telescopes were mounted on astronomical quadrants (p.12), and to stabilize the telescope, the quadrant was usually mounted on a wall. These kinds of telescopes are called mural quadrants from the Latin word for "wall", *murus*. The telescope was hung on a single pivot-point, so that its eyepiece could be moved along the graduated scale of the arc of the quadrant (p.12). In this way, astronomers could accurately measure the altitude of the stars they were observing.

GEMINI TELESCOPE

There are two Gemini Telescopes, one on Hawaii (in the northern hemisphere) and one in Chile (in the southern hemisphere). Together they give optical and infrared coverage of the whole sky. Each Gemini Telescope has a single active mirror that is 8.1 m (26.6 ft) across. The mirrors have protected silver coatings that help to prevent interference in the infrared spectrum.

GRINDING MIRRORS

The 5 m (16 ft) mirror of the famous Hale telescope on Mount Palomar in California, US was cast in 1934 from 35 tonnes of molten Pyrex. The grinding of the mirror to achieve the correct curved shape was interrupted by the Second World War. It was not completed until 1947. Mount Palomar was one of the first high-altitude observatories, built where the atmosphere is thinner and the effects of pollution are reduced.

A SEGMENTED-MIRROR TELESCOPE

Inside each of the twin Keck Telescopes on Hawaii, there is a primary six-sided mirror that is around 10 m (33 ft) wide. It is made up of 36 smaller hexagonal mirrors, which are 1.8 m (6 ft) across. Each small mirror is monitored by a computer and its position can be adjusted to correct any sagging. The two telescopes are also linked so that they can combine their signals for an even more accurate image.

Observatories

AN OBSERVATORY IS A PLACE where astronomers watch the heavens. The shapes of observatories have changed greatly over the ages (p.8). The earliest were quiet places set atop city walls or in towers. Height was important so that the astronomer could have a panoramic, 360° view of the horizon. The Babylonians and the Greeks certainly had rudimentary observatories, but the greatest of the early observatories were those in Islamic North Africa and the Middle East – Baghdad, Cairo, and Damascus. The great observatory at Baghdad had a huge 6 m (20 ft) quadrant and a 17 m (56 ft) stone sextant. It must have looked very much like the observatory at Jaipur – the only one of this type of observatory to remain relatively intact (below). As the great Islamic empires waned and science reawakened in western Europe, observatories took on a different shape. The oldest observatory still in use is the Observatoire de Paris, founded in 1667 (p.28). A less hospitable climate meant that open-air observatories were impractical. The astronomer and the instruments needed a roof over their heads. Initially, these roofs were constructed with sliding panels or doors that could be pulled back to open the building to the night sky. Since the 19th century most large telescopes are covered with huge rotatable domes. The earliest domes were made of papier mâché, the only substance known to be sufficiently light and strong. Now most domes are made of aluminium.

THE LEVIATHAN OF PARSONSTOWN
William Parsons (1800–1867), the third Earl of Rosse, was determined to build the largest reflecting telescope. At Parsonstown in Ireland he managed to cast a 182-cm (72-in) mirror, weighing nearly 4 tonnes and magnifying 800–1,000 times. When the "Leviathan" was built in 1845, it was used by Parsons to make significant discoveries concerning the structure of galaxies and nebulae (pp.60–63).

BEIJING OBSERVATORY
The Great Observatory set on the walls of the Forbidden City in Beijing, China, was constructed with the help of Jesuit priests from Portugal in 1660 on the site of an older observatory. The instruments included two great armillary spheres (p.11), a huge celestial globe (p.10), a graduated azimuth horizon ring, and an astronomical quadrant and sextant (p.12). The shapes of these instruments were copied from woodcut illustrations in Tycho Brahe's *Mechanica* of 1598 (p.18).

JAIPUR, INDIA
Early observations were carried out by the naked eye from the top of monumental architectural structures. The observatory at Jaipur in Rajasthan, India, was built by Maharajah Jai Singh in 1726. The monuments include a massive sundial, the Samrat Yantra, and a gnomon inclined at 27°, showing the latitude of Jaipur and the height of the Pole Star (p.13). There is also a large astronomical sextant and a meridian chamber.

MAUNA KEA

Increasing use of artificial light and air pollution from the world's populous cities have driven astronomers to the most uninhabited regions of the Earth to build their observatories. The best places are mountain tops or deserts in temperate climates where the air is dry, stable and without clouds. The Mauna Kea volcano on the island of Hawaii has the thinner air of high altitudes and the temperate climate of the Pacific. There are optical, infrared, and radio telescopes here.

COMPUTER-DRIVEN TELESCOPE

Telescopes have become so big that astronomers are dwarfed by them. This 51-cm (20-in) solar coronagraph in the Crimean Astrophysical Observatory in the Ukraine is driven by computer-monitored engines. A coronagraph is a type of solar telescope that measures the outermost layers of the Sun's atmosphere (p.38).

What is a meridian?

Meridian lines are imaginary coordinates running from pole to pole that are used to measure distances east and west on the Earth's surface and in the heavens. Meridian lines are also known as lines of longitude. The word meridian comes from the Latin word *meridies*, meaning "the mid-day", because the Sun crosses a local meridian at noon. Certain meridians became important because astronomers used them in observatories when they set up their telescopes for positional astronomy. This means that all their measurements of the sky and the Earth were made relative to their local meridian. Until the end of the 19th century there were a number of national meridians in observatories in Paris, Cadiz, and Naples.

Prime meridian

THE GREENWICH MERIDIAN

In 1884 there was an international conference in Washington DC to agree a single Zero Meridian, or Prime Meridian for the world. The meridian running through the Airy Transit Circle – a telescope mounted so that it rotated in a north/south plane – at the Royal Greenwich Observatory outside London was chosen. This choice was largely a matter of convenience. Most of the shipping charts and all of the American railway system used Greenwich as their longitude zero at the time. South of Greenwich the Prime Meridian crosses through France and Africa, and then runs across the Atlantic Ocean all the way to the South Pole.

CROSSING THE MERIDIAN

In 1850 the 7th Astronomer Royal of Great Britain, Sir George Biddle Airy (1801–1892), decided he wanted a new telescope. In building it, he moved the previous Prime Meridian for England 5.75 m (19 ft) to the east. The Greenwich Meridian is marked by a green laser beam projected into the sky and by an illuminated line which bisects Airy's Transit Circle at the Royal Observatory.

Astronomers

FASHIONABLE AMATEURS
By the 18th century the science of the stars became an acceptable pastime for the rich and sophisticated. The large number of small telescopes that survive from this period is evidence of how popular amateur astronomy had become.

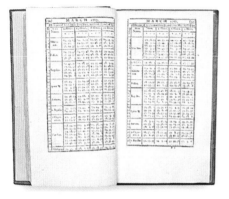

THE NAUTICAL ALMANAC
First published in 1766, *The Nautical Almanac* provides a series of tables showing the distances between certain key stars and the Moon at three-hourly intervals. Navigators can use the tables to help calculate their longitude at sea, when they are out of sight of land (p.27).

The main difference between astronomers and most other scientists is that astronomers can only conduct direct experiments in the solar system – by sending spacecraft. They cannot experiment on stars and galaxies. The key to most astronomy is careful and systematic observing. Astronomers must watch and wait for things to happen. Early astronomers could do little more than plot the positions of the heavenly bodies, follow their movements in the sky and be alert for unexpected events, such as the arrival of a comet. From the 19th century, astronomers began to investigate the physics of the universe by analysing light and other radiations from space. But the sorts of questions astrophysicists still try to answer today are very similar to the questions that puzzled the earliest Greek philosophers – what is the Universe, how is it shaped, and how do I fit into it?

FIRST ASTRONOMER ROYAL
England appointed its first Astronomer Royal, John Flamsteed (1646–1719), in 1675. He lived and worked at the Royal Observatory, Greenwich, built by King Charles II of England in the same year.

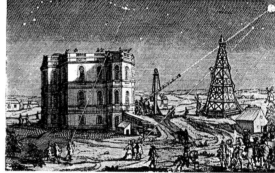

IN THE FAMILY
When the Observatoire de Paris was founded in 1667, the French King Louis XIV called a well-known Bolognese astronomer, Gian Domenico Cassini (1625–1712), to Paris to be the observatory's director. He was followed by three generations of Cassinis in the position: Jacques Cassini (1677–1756); César–François Cassini de Thury (1714–1784), who produced the first modern map of France; and Jean–Dominique Cassini (1748–1845). Most historians refer to this great succession of astronomers simply as Cassini I, Cassini II, Cassini III, and Cassini IV.

ASTRONOMY IN RUSSIA
The Russian astronomer Mikhail Lomonosov (1711–1765) was primarily interested in problems relating to the art of navigation and fixing latitude and longitude. During his observations of the 1761 Transit of Venus (pp.46–47), he noticed that the planet seemed "smudgy", and suggested that Venus had a thick atmosphere, many times denser than that of the Earth.

STAR CLOCK (1815)
One of the primary aspects of positional astronomy is measuring a star's position against a clock.
This ingenious clock has the major stars inscribed on the surface of its rotating face. Placing pegs in the holes near the stars to be observed causes the clock to chime when the star is due to pass the local meridian.

Peg marking α Cassiopeiae   *Peg marking α Aquarii*   *Rotating clock face*

Peg marking Antares

Peg marking α Hydrae

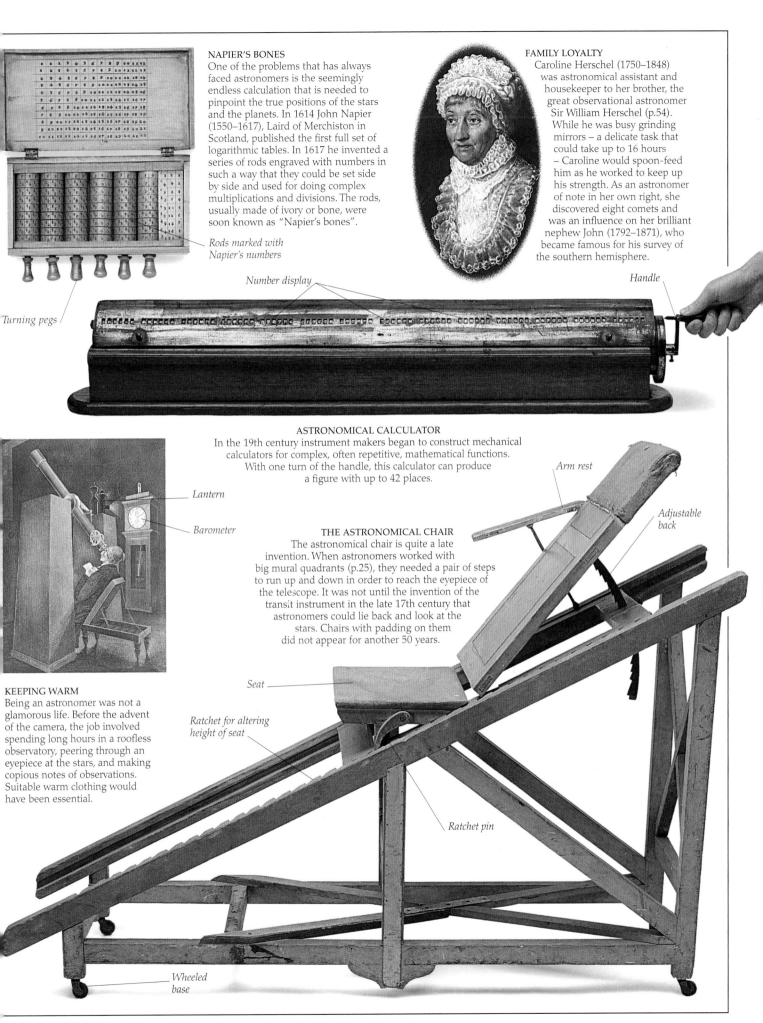

NAPIER'S BONES

One of the problems that has always faced astronomers is the seemingly endless calculation that is needed to pinpoint the true positions of the stars and the planets. In 1614 John Napier (1550–1617), Laird of Merchiston in Scotland, published the first full set of logarithmic tables. In 1617 he invented a series of rods engraved with numbers in such a way that they could be set side by side and used for doing complex multiplications and divisions. The rods, usually made of ivory or bone, were soon known as "Napier's bones".

Rods marked with Napier's numbers

FAMILY LOYALTY

Caroline Herschel (1750–1848) was astronomical assistant and housekeeper to her brother, the great observational astronomer Sir William Herschel (p.54). While he was busy grinding mirrors – a delicate task that could take up to 16 hours – Caroline would spoon-feed him as he worked to keep up his strength. As an astronomer of note in her own right, she discovered eight comets and was an influence on her brilliant nephew John (1792–1871), who became famous for his survey of the southern hemisphere.

Turning pegs

Number display

Handle

ASTRONOMICAL CALCULATOR

In the 19th century instrument makers began to construct mechanical calculators for complex, often repetitive, mathematical functions. With one turn of the handle, this calculator can produce a figure with up to 42 places.

Lantern

Barometer

Arm rest

Adjustable back

THE ASTRONOMICAL CHAIR

The astronomical chair is quite a late invention. When astronomers worked with big mural quadrants (p.25), they needed a pair of steps to run up and down in order to reach the eyepiece of the telescope. It was not until the invention of the transit instrument in the late 17th century that astronomers could lie back and look at the stars. Chairs with padding on them did not appear for another 50 years.

Seat

KEEPING WARM

Being an astronomer was not a glamorous life. Before the advent of the camera, the job involved spending long hours in a roofless observatory, peering through an eyepiece at the stars, and making copious notes of observations. Suitable warm clothing would have been essential.

Ratchet for altering height of seat

Ratchet pin

Wheeled base

Spectroscopy

ASTRONOMERS HAVE BEEN ABLE to study the chemical composition of the stars and how hot they are for more than a century by means of spectroscopy. A spectroscope breaks down the "white" light coming from a celestial body into an extremely detailed spectrum. Working on Isaac Newton's discovery of the spectrum (p.21), a German optician, Josef Fraunhofer (1787–1826), examined the spectrum created by light coming from the Sun and noticed a number of dark lines crossing it. In 1859 another German, Gustav Kirchhoff (1824–1887), discovered the significance of Fraunhofer's lines. They are produced by chemicals in the cooler, upper layers of the Sun (or a star) absorbing light. Each chemical has its own pattern of lines, like a fingerprint. By looking at the spectrum of the Sun, astronomers have found all the elements that are known on the Earth in the Sun's atmosphere.

THE COLOURS OF THE RAINBOW
A rainbow is formed by the Sun shining through raindrops. The light is refracted by droplets of water as if each one were a prism.

Prism splits the light into its colours

The spectrum

Infrared band

Red

Rays of white light

Violet

Sodium lamp

The spectroscope would be mounted on a telescope here

HERSCHEL DISCOVERS INFRARED
In 1800 Sir William Herschel (p.54) set up a number of experiments to test the relationship between heat and light. He repeated Newton's experiment of splitting white light into a spectrum (p.21) and, by masking all the colours but one, was able to measure the individual temperatures of each colour in the spectrum. He discovered that the red end of the spectrum was hotter than the violet end, but was surprised to note that an area where he could see no colour, next to the red end of the spectrum, was much hotter than the rest of the spectrum. He called this area infrared or 'below the red'.

Stand for photographic plate

Diffraction grating

Solar spectrum showing absorption lines

Sodium

Spectroscope

Emission spectrum of sodium *Sodium*

LOOKING AT SODIUM
Viewing a sodium flame through a spectroscope can help to explain how spectroscopy works in space. According to Gustav Kirchhoff's first law of spectral analysis, a hot dense gas at high pressure produces a continuous spectrum of all colours. His second law states that a hot rarefied gas at low pressure produces an emission line spectrum, characterized by bright spectral lines against a dark background. His third law states that when light from a hot dense gas passes through a cooler gas before it is viewed, it produces an absorption line spectrum – a bright spectrum riddled with a number of dark, fine lines.

WHAT IS IN THE SUN?
When a sodium flame is viewed through a spectroscope (left), the emission spectrum produces the characteristic bright yellow lines (above). The section of the Sun's spectrum (top) shows a number of tiny "gaps" or dark lines. These are the Fraunhofer lines from which the chemical composition of the Sun can be determined. The two dark lines in the yellow part of the spectrum correspond to the sodium. As there is no sodium in the Earth's atmosphere, it must be coming from the Sun.

KIRCHHOFF AND BUNSEN
Following the invention of the clean-flame burner by the German chemist Robert Bunsen (1811–1899) it was possible to study the effect of different chemical vapours on the known pattern of spectral lines. Together, Gustav Kirchhoff and Bunsen invented a new instrument called the spectroscope to measure these effects. Within a few years, they had managed to isolate the spectra for many known substances, as well as to discover a few unknown elements.

Continuous spectrum

ABSORBING COLOUR
To prove his laws of spectral analysis, Kirchhoff used sodium gas to show that when white light is directed through the gas, the characteristic colour of the sodium is absorbed and the spectrum shows black lines where the sodium should have appeared. In the experiment shown above, a continuous spectrum (top) is produced by shining white light through a lens. When a petri dish of the chemical potassium permanganate in solution is placed between the lens and the light, some of the colour of the spectrum is absorbed.

The spectrum of potassium permanganate

SPECTRUM OF THE STARS
By closely examining the spectral lines in the light received from a distant star, the astronomer can detect these "fingerprints" and uncover the chemical composition of the object being viewed. Furthermore, the heat of the source can also be discovered by studying the spectral lines. Temperature can be measured by the intensities of individual lines in their spectra. The width of the line provides information about temperature, movement, and presence of magnetic fields. With magnification, each of these spectra can be analyzed in more detail.

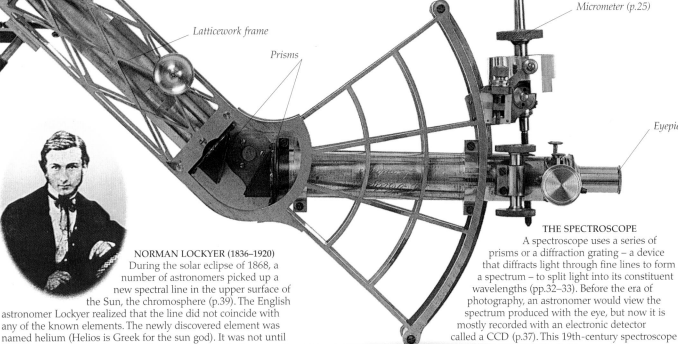

Eyepiece

Micrometer (p.25)

Latticework frame

Prisms

Eyepiece

NORMAN LOCKYER (1836–1920)
During the solar eclipse of 1868, a number of astronomers picked up a new spectral line in the upper surface of the Sun, the chromosphere (p.39). The English astronomer Lockyer realized that the line did not coincide with any of the known elements. The newly discovered element was named helium (Helios is Greek for the sun god). It was not until 1895, however, that helium was discovered on the Earth.

THE SPECTROSCOPE
A spectroscope uses a series of prisms or a diffraction grating – a device that diffracts light through fine lines to form a spectrum – to split light into its constituent wavelengths (pp.32–33). Before the era of photography, an astronomer would view the spectrum produced with the eye, but now it is mostly recorded with an electronic detector called a CCD (p.37). This 19th-century spectroscope uses a prism to split the light.

The radio telescope

WITH THE DISCOVERY OF non-visible light, such as infrared (p.30), and electromagnetic and X-ray radiations, scientists began to wonder if objects in space might emit invisible radiation as well. The first such radiation to be discovered (by accident) was radio waves – the longest wavelengths of the electromagnetic spectrum. To detect radio waves, astronomers constructed huge dishes in order to capture the long waves and "see" detail. Even so, early radio telescopes were never large enough, proportionally, to catch the fine features that optical telescopes could resolve. Today, by electronically combining the output from many radio telescopes, a dish the size of the Earth can be synthesized, revealing details many times finer than optical telescopes. Astronomers routinely study all radiations from objects in space, often using detectors high above the Earth's atmosphere (p.7).

RADIO GALAXY
This image shows the radio emission from huge invisible clouds of very hot gas beamed out from a black hole in the centre of a galaxy called NGC 1316. The maps of the radio clouds, shown in orange, were made by the Very Large Array (p.33).

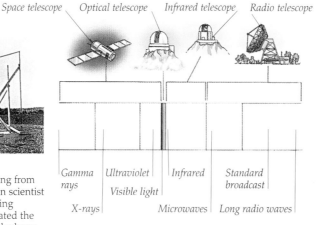

Space telescope   Optical telescope   Infrared telescope   Radio telescope

Gamma rays   *Ultraviolet*   *Infrared*   *Standard broadcast*

Visible light

X-rays   *Microwaves*   *Long radio waves*

ELECTROMAGNETIC SPECTRUM
The range of frequencies of electromagnetic radiation is known as the electromagnetic spectrum. Very low on the scale are radio waves, rising to infrared (p.30), visible light, ultraviolet, and X-rays, with gamma rays at the highest frequency end of the spectrum. The radiations that pass through the Earth's atmosphere are light and radio waves, though infrared penetrates to the highest mountaintops. The remainder can only be detected by sending instruments into space (pp.34–35). All telescopes – radio, optical, and infrared – "see" different aspects of the sky, caused by the different physical processes going on.

EVIDENCE OF RADIO RADIATION
The first evidence of radio radiation coming from outer space was collected by the American scientist Karl Jansky (1905–1950) who, in 1931, using home-made equipment (above), investigated the static affecting short-wavelength radio-telephone communication. He deduced that this static must be coming from the centre of our galaxy (pp.62–63).

AMATEUR ASTRONOMER
On hearing about Jansky's discoveries, American amateur astronomer Grote Reber (1911–2002) built a large, moveable radio receiver in his backyard in 1936. It had a parabolic surface to collect the radio waves. With this 9 m (29 ft) dish, he began to map the radio emissions coming from the Milky Way. For years Reber was the only radio astronomer in the world.

ARECIBO TELESCOPE
The mammoth Arecibo radio dish is built in a natural limestone concavity in the jungle south of Arecibo, Puerto Rico. The "dish", which is a huge web of steel mesh, measures 305 m (1,000 ft) across, providing a 20-hectare (50-acre) collecting surface. Although the dish is fixed, overhead antennae can be moved to different parts of the sky.

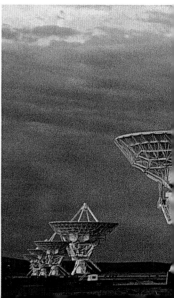

HOT SPOTS

Radio astronomers can create temperature maps of planets. This false-colour map shows temperatures just below Mercury's surface. Because Mercury is so close to the Sun, the hottest area is on Mercury's equator, shown here as red. The blue areas are the coolest.

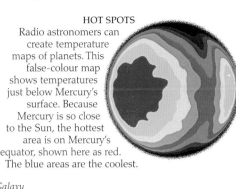

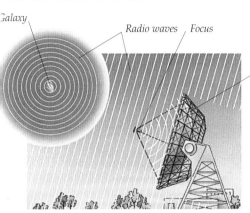

Galaxy   *Radio waves*   *Focus*

Parabolic dish

BERNARD LOVELL

The English astronomer, Bernard Lovell (b. 1913), was a pioneer of radio astronomy. He developed a research station at Jodrell Bank, England, in 1945 using surplus army radar equipment. He is seen here in the control room of the 76-m (250-ft) diameter Mark 1 radio telescope (later renamed the Lovell Telescope in his honour). The telescope's giant dish was commissioned in 1957.

HIGH-TECH TELESCOPE

Communications technology allows astronomers to work nearly anywhere in the world. All they need is a computer link-up. While optical telescopes are sited far from built-up areas (p.27), clear skies are not necessary for radio astronomy. This telescope is the world's largest, fully steerable, single-dish radio telescope; it is 100 m (330 ft) in diameter and is located near Bonn, Germany.

HOW A RADIO TELESCOPE WORKS

The parabolic dish of a radio telescope can be steered to pick up radio signals. It focuses them to a point from which they are sent to a receiver, a recorder, and then a data room at a control centre. Computer equipment then converts intensities of the incoming radio waves into images that are recognizable to our eyes as objects from space (p.57).

A VERY LARGE ARRAY

Scientists soon realized that radio telescopes could be connected together to form very large receiving surfaces. For example, two dishes 100 km (62 miles) apart can be linked electronically so that their receiving area is the equivalent of a 100-km (62-mile) wide dish. One of the largest arrangements of telescopes is the Very Large Array (VLA) set up in the desert near Socorro, New Mexico, USA. Twenty-seven parabolic dishes have been arranged in a huge "Y", covering more than 27 km (17 miles).

Parabolic dish

Mounting and support

Venturing into space

S**INCE THE LAST APOLLO MISSION** to the Moon in 1972, no human has travelled any further into space than Earth orbit. But the exploration and exploitation of space have not stopped. Dozens of spacecraft carrying instruments and cameras have travelled way beyond the Moon to investigate planets and moons, asteroids and comets, the Sun and interplanetary space. Instead of competing, countries collaborate and share costs. Space science and technology bring huge benefits to our lives. TV services use orbiting communications satellites. Ships, aircraft and road traffic navigate using satellite signals. Military satellites are used for surveillance. Weather forecasts use images from meteorological satellites and resources satellites gather detailed information about the Earth's surface. And NASA is now planning to send more astronauts to the Moon by 2020. They will set up a lunar base for research and for testing the technologies needed to send humans to Mars.

Luna 1

LUNAR PROBES
The former USSR launched *Sputnik 1*, the first artificial satellite, into space in 1957. Between the late 1950s and 1976 several probes were sent to explore the surface of the Moon. *Luna 1* was the first successful lunar probe. It passed within 6,000 km (3,730 miles) of the Moon. *Luna 3* was the first probe to send back pictures to the Earth of the far side of the Moon (pp.40–41). The first to achieve a soft landing was *Luna 9* in February 1966. *Luna 16* collected soil samples, bringing them back without any human involvement. The success of these missions forced people to take space exploration more seriously.

GETTING INTO SPACE
The American physicist Robert Goddard (1882–1945) launched the first liquid-fuelled rocket in 1926. This fuel system overcame the major obstacle to launching an orbiting satellite which was the weight of solid fuels. If a rocket is to reach a speed great enough to escape the Earth's gravitational field, it needs a thrust greater than the weight it is carrying.

THE FIRST MAN IN SPACE
On 12 April 1961 the former USSR (now Russia) launched the 5-tonne spaceship *Vostok 1*. It was manned by the cosmonaut Yuri Gagarin (1934–1968) who made a complete circuit of the Earth at a height of 303 km (188 miles). He remained in space for 1 hour and 29 minutes before landing back safely in the USSR. He was hailed as a national hero and is seen here being lauded by the Premier of the USSR, Nikita Khrushchev.

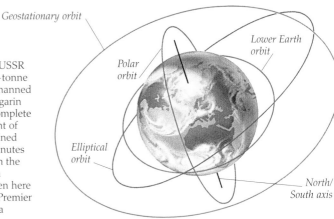

Geostationary orbit

Lower Earth orbit

Polar orbit

Elliptical orbit

North/ South axis

SATELLITE ORBITS
A satellite is sent into an orbit that is most suitable for the kind of work it has to do. Space telescopes such as Hubble (p.7), take the low orbits – 600 km (373 miles) above the Earth's surface. US spy and surveillance satellites orbit on a north/south axis to get a view of the whole Earth, while those belonging to Russia often follow elliptical orbits that allow them to spend more time over their own territory. Communications and weather satellites are positioned above the equator. They take exactly 24 hours to complete an orbit, and therefore seem to hover above the same point on the Earth's surface – known as a geostationary orbit.

LUNAR LANDING
Between 1969 and 1972 six manned lunar landings took place. The first astronaut to set foot on the Moon was Neil Armstrong (b. 1930) on 21 July 1969. Scientifically, one of the major reasons for Moon landings was to try to understand the origin of the Moon itself and to understand its history and evolution. This photograph shows American astronaut James Irwin with the *Apollo 15* Lunar Rover in 1971.

COOPERATION IN SPACE

The European Space Agency (ESA) is an organisation through which 16 European countries collaborate on a joint space programme. It provides the means for a group of smaller countries to participate in space exploration and share the benefits of space-age technology. ESA has its own rocket, called *Ariane*, which is launched from a space-port in French Guiana. In 2003, this *Ariane 5* rocket launched the *SMART-1* spacecraft on a mission to orbit the Moon and to test a new spacecraft propulsion technology. In addition to the USA and Russia, several other major countries have their own space agencies, including Japan and China.

LIVING IN SPACE

Construction of the International Space Station (ISS) began in 1998 and continues until 2010. It is a joint project between the USA, Europe, Russia, Canada, and Japan. The ten main modules and other parts are being transported by the Space Shuttle or by an unmanned Russian space vehicle. The first crew arrived in 2000 and there have been at least two astronauts on board ever since. The ISS takes 92 minutes to orbit Earth at an average height of 354 km (220 miles).

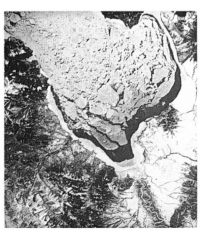

BENEFITS OF SATELLITES

Meteorological satellites can monitor the changing patterns of the weather and plot ocean currents, which play a major role in determining the Earth's climate. Data gathered by monitoring such vast expanses as this Russian ice floe can be used to predict climate change. Resource satellites are used for geological and ecological research. For example, they map the distribution of plankton – a major part of the food chain – in ocean waters.

External fuel tank

The Space Shuttle

The first flight of a Space Shuttle was in 1981. Since then, five Shuttles have made a total of over 120 flights into Earth orbit. Their tasks have included launching satellites, repairing the Hubble Space Telescope, and taking parts and crew to the International Space Station. Two of the Shuttles have been destroyed in accidents and the others will go out of service in 2010.

THE SPACE SHUTTLE

The Shuttle is boosted into space by two huge, reusable, solid-fuel booster rockets. They are jettisoned and then fall back to Earth slowed by parachutes so they can be retrieved. The Shuttle returns to Earth and lands at about 350 km/h (217 mph). It is protected from the intense heat of re-entry by a shell of thermal tiles.

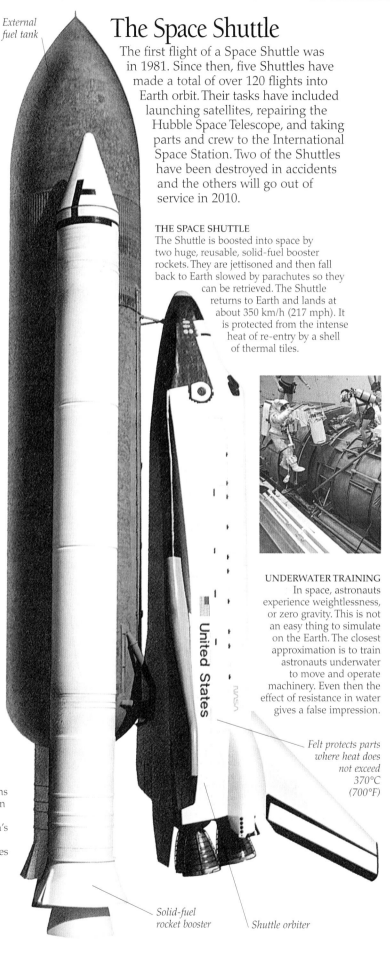

UNDERWATER TRAINING

In space, astronauts experience weightlessness, or zero gravity. This is not an easy thing to simulate on the Earth. The closest approximation is to train astronauts underwater to move and operate machinery. Even then the effect of resistance in water gives a false impression.

Felt protects parts where heat does not exceed 370°C (700°F)

Solid-fuel rocket booster

Shuttle orbiter

The Solar System

THE SOLAR SYSTEM is the group of planets, moons, and space debris orbiting around our Sun. It is held together by the gravitational pull of the Sun, which is nearly 1,000 times more massive than all the planets put together. The Solar System was probably formed from a huge cloud of interstellar gas and dust that contracted under the force of its own gravity five billion years ago. The planets are divided into two groups. The four planets closest to the Sun are called "terrestrial" from the Latin word *terra* meaning "land" because they are small and dense and have hard surfaces. The four outer planets are called "Jovian" because, like Jupiter, they are giant planets made largely of gas and liquid. Between Mars and Jupiter and beyond Neptune there are belts of very small bodies and dwarf planets called the asteroid belt and the Kuiper belt.

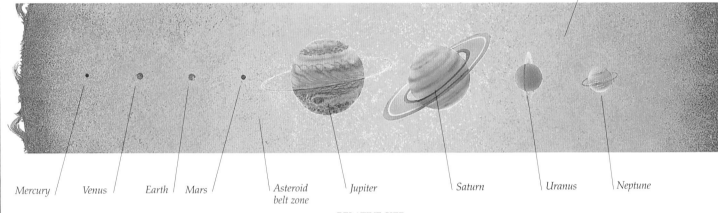

Sun

Mercury | Venus | Earth | Mars | Asteroid belt zone | Jupiter | Saturn | Uranus | Neptune

RELATIVE SIZE

The Sun has a diameter of approximately 1,392,000 km (865,000 miles). It is almost ten times larger than the largest planet, Jupiter, which is itself big enough to contain all the other planets put together. The planets are shown here to scale against the Sun. Those planets with orbits inside the Earth are sometimes referred to as the inferior planets; those beyond the Earth are the superior planets. The four small planets that orbit the Sun relatively closely – Mercury, Venus, Earth, and Mars – have masses lower than those of the next four, but have much greater densities (p.45). Jupiter, Saturn, Uranus, and Neptune have large masses with low densities. They are more widely spaced apart and travel at great distances from the Sun.

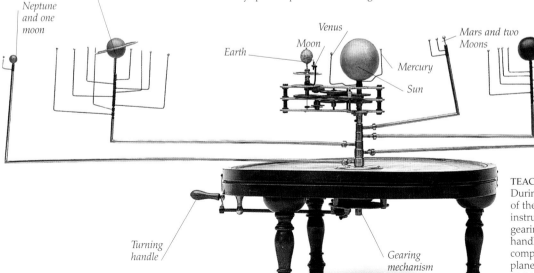

Saturn and eight moons

Neptune and one moon

Venus

Earth

Moon

Mercury

Sun

Mars and two Moons

Jupiter and nine moons

Uranus and four moons

Turning handle

Gearing mechanism

TEACHING ASTRONOMY
During the 19th century, the astronomy of the Solar System was taught by mechanical instruments such as this orrery. The complex gearing of the machine is operated by a crank handle, which ensures that each planet completes its solar orbit relative to the other planets. The planets are roughly to a scale of 80,500 km (50,000 miles) to 3 cm (1 in), except for the Sun, which would need to be 43 cm (17 in) in diameter for the model to be accura

CELESTIAL MECHANICS

The Frenchman Pierre Simon Laplace (1749–1827) was the first scientist to make an attempt to compute all the motions of the Moon and the planets by mathematical means. In his five-volume work, *Traité de méchanique céleste* (1799–1825), Laplace treated all motion in the Solar System as a purely mathematical problem, using his work to support the theory of universal gravitation (p.21). His idea, for which he was severely criticized during the following century, was that the heavens were a great celestial machine, like a timepiece that, once set in motion, would go on forever.

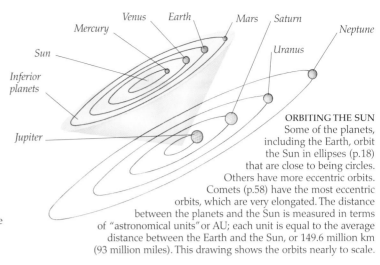

Sun
Mercury *Venus* *Earth* *Mars* *Saturn* *Uranus* *Neptune*
Inferior planets
Jupiter

ORBITING THE SUN

Some of the planets, including the Earth, orbit the Sun in ellipses (p.18) that are close to being circles. Others have more eccentric orbits. Comets (p.58) have the most eccentric orbits, which are very elongated. The distance between the planets and the Sun is measured in terms of "astronomical units" or AU; each unit is equal to the average distance between the Earth and the Sun, or 149.6 million km (93 million miles). This drawing shows the orbits nearly to scale.

Photographing the planets

One of the key tasks of space missions (pp.34–35) is to send back pictures of distant planets and moons. They do this using imaging devices very similar to those used in digital cameras. The heart of the system is a CCD, or charge-coupled device. This is a silicon chip with thousands of light-sensitive pixels, or picture elements. The amount of light falling on each pixel produces a different electrical signal. This is read by an on-board computer and converted into a stream of digital signals that can be radioed back to Earth where they are reconstructed into the image by computer.

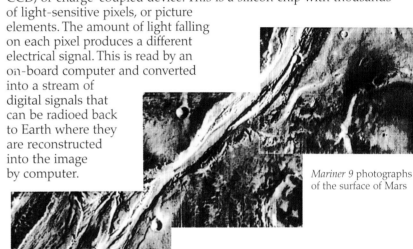

Mariner 9 photographs of the surface of Mars

Escaping elements

Lighter elements

By increasing the vibration, the balls are given more energy

Heavier elements

Kinetic energy machine

CREATING COLOUR

The CCDs used in astronomy rarely produce colour images directly, but use the most sensitive black-and-white chips. To get a colour image separate images are taken through colour filters, and the results are combined in a computer to give a realistic colour view.

HYDROGEN IN THE SOLAR SYSTEM

Hydrogen is a common element in the Solar System. Hydrogen atoms are so energetic that lightweight planets cannot hang on to them. This is why the heavier nitrogen makes up such a high percentage of the Earth's atmosphere (p.42). Lighter hydrogen has escaped because the Earth's gravity is not strong enough to hold on to it. The red balls in this kinetic energy machine represent the heavier elements; the tiny silver balls represent the lighter elements such as hydrogen. Our massive Sun is made up largely of hydrogen. Its great mass pulls the hydrogen inwards and, at its core, hydrogen fuses into helium under the extreme heat and pressure. It is this reaction, like a giant hydrogen bomb, that makes the Sun shine. Hydrogen also makes up a large part of Jupiter, Saturn, Uranus, and Neptune (pp.50–57).

COLOUR MOSAIC OF MARS

The detail in an individual CCD image of a planet is limited by the number of pixels on the chip. To get a high-quality image, several shots are taken of different parts of the planet, and then a mosaic is produced like this one of Mars.

The Sun

Aʟᴍᴏsᴛ ᴇᴠᴇʀʏ ᴀɴᴄɪᴇɴᴛ ᴄᴜʟᴛᴜʀᴇ recognized the Sun as the giver of life and primary power behind events here on Earth. The Sun is the centre of our Solar System, our local star. It has no permanent features because it is gaseous – mainly incandescent hydrogen. The temperature of the Sun's visible yellow disk – the photosphere – is about 5500°C (9900°F). Over the photosphere, there are layers of hotter gas – the chromosphere and corona. The thin gas in the corona is at about a million degrees. By using spectroscopic analysis (pp.30–31), scientists know that the Sun, like most stars (pp.60–61), is made up mostly of hydrogen. In its core, the hydrogen nuclei are so compressed that they eventually fuse into helium. This is the same thing that happens in a hydrogen atomic bomb. Every minute, the Sun converts 240 million tonnes of mass into energy. Albert Einstein's famous formula, $E=mc^2$, shows how mass and energy are mutually interchangeable (p.63), helping scientists to understand the source of the Sun's energy.

THE CORONAGRAPH
In 1930 the French astronomer Bernard Lyot (1897–1952) invented the coronagraph. It allows the Sun's corona to be viewed without waiting for a total solar eclipse.

VIEWING THE SUN
Even though the Sun is more than 149 million km (93 million miles) from the Earth, its rays are still bright enough to damage the eyes permanently. The Sun should *never* be viewed directly and certainly not through a telescope or binoculars. Galileo went blind looking at the Sun. This astronomer is at the Kitt Peak National Observatory in Arizona, US. Two mirrors at the top of the solar telescope tower reflect the Sun's image down a tube to the mirror below. Inside the tube there is a vacuum. This prevents distortion that would be caused by the air in the tower.

THE DIPLEIDOSCOPE
Local noon occurs when the Sun crosses the local north/south meridian (p.27). In the 19th century a more accurate device than the gnomon (p.14) was sought to indicate when noon occurred. The dipleidoscope, invented in 1842, is an instrument with a hollow, right-angled prism, which has two silvered sides and one clear side. As the Sun passes directly overhead, the two reflected images are resolved into a single one. This shows when it is local noon.

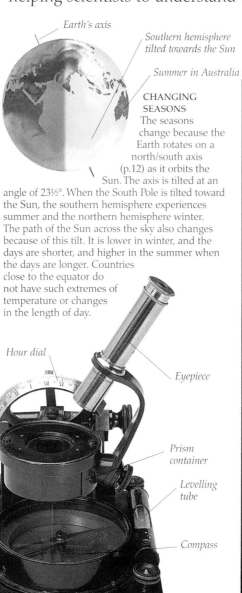

Earth's axis

Southern hemisphere tilted towards the Sun

Summer in Australia

CHANGING SEASONS
The seasons change because the Earth rotates on a north/south axis (p.12) as it orbits the Sun. The axis is tilted at an angle of 23½°. When the South Pole is tilted toward the Sun, the southern hemisphere experiences summer and the northern hemisphere winter. The path of the Sun across the sky also changes because of this tilt. It is lower in winter, and the days are shorter, and higher in the summer when the days are longer. Countries close to the equator do not have such extremes of temperature or changes in the length of day.

Hour dial

Eyepiece

Prism container

Levelling tube

Compass

Chromosphere

THE CORONA

The outermost layer of the Sun's atmosphere is called the corona. Even though it extends millions of kilometres into space, it cannot be seen during the day because of the brightness of the blue sky. During a total eclipse, the corona appears like a crown around the Moon. It is clearly seen in this picture of a total eclipse over Mexico in March 1970.

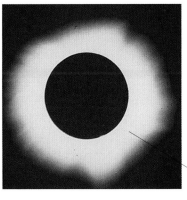

Corona

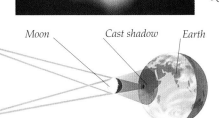

Sun   *Moon*   *Cast shadow*   *Earth*

SOLAR ECLIPSE

A solar eclipse happens when a new Moon passes directly between the Earth and the Sun, casting a shadow on the surface of the Earth. From an earthly perspective, it looks as if the Moon has blocked out the light of the Sun. Total eclipses of the Sun are very rare in any given location, occurring roughly once every 360 years in the same place. However, several solar eclipses may occur each year.

Sunspots

Sunspots are cooler areas on the Sun, where strong magnetic fields disturb the flow of heat from the core to the photosphere. Typical sunspots last about a week and are twice as big as Earth. They often form in pairs or groups. The number of sunspots appearing on the Sun rises and falls over an 11-year period. This is called the solar cycle. At sunspot maximum the Sun also experiences large explosive eruptions called flares, which blast streams of particles into space.

PLOTTING THE SUNSPOTS

By observing the changing position of sunspots, we can see that the Sun is spinning. Unlike the planets, however, the whole mass of the Sun does not spin at the same rate because it is not solid. The Sun's equator takes 25 Earth days to make one complete rotation. The Sun's poles take nearly 30 days to accomplish the same task. These photographs are a record of the movements of a large spot group on 14 days in March/April 1947.

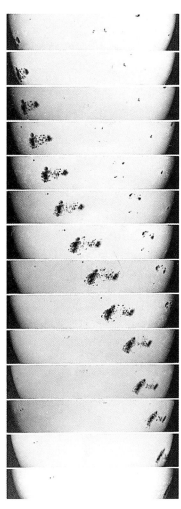

Prominence

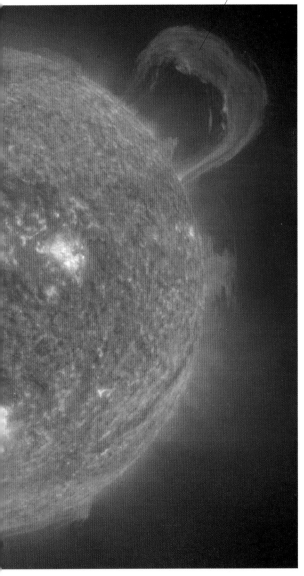

CORONAL LOOPS

Huge loops of very hot gas surge through the Sun's corona, guided by the magnetic field. These loops are about 30 times larger than Earth. This picture was taken from space in extreme ultraviolet light by NASA's TRACE satellite, launched in 1998 to study the Sun.

SOLAR PROMINENCE

Astronomers have learned much about the Sun from solar observatories operating in space, such as SOHO (the Solar and Heliospheric Observatory). This SOHO image of the Sun shows ultraviolet light from the chromosphere, a layer of hot gas above the yellow disk of the Sun we normally see. A huge prominence is erupting into the corona. Prominences like this usually last a few hours. They can fall back down or break off and cause gas to stream into space. Sometimes, the corona blasts huge clouds of gas into space. If one of these coronal mass ejections reaches Earth, it may cause a magnetic storm and trigger an aurora (northern or southern lights).

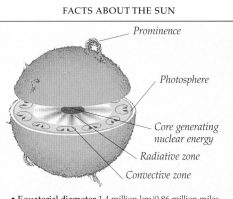

FACTS ABOUT THE SUN

Prominence

Photosphere

Core generating nuclear energy

Radiative zone

Convective zone

- **Equatorial diameter** 1.4 million km/0.86 million miles
- **Distance from the Earth** 149 million km/ 93 million miles
- **Rotational period** 25 Earth days
- **Volume** (Earth =1) 1,306,000
- **Mass** (Earth =1) 333,000
- **Density** (water = 1) 1.41
- **Temperature at surface** 5,500°C

The Moon

THE MOON IS THE EARTH'S only satellite, about 384,000 km (239,000 miles) away. Next to the Sun it is the brightest object in our sky, more than 2,000 times as bright as Venus. Even without a telescope, we can see large areas on the Moon that are darker than the rest. Early observers imagined these might be seas and they were given names such as the Sea of Tranquillity. We now know that there is neither liquid water nor an atmosphere on the Moon. The so-called "seas" are plains of volcanic rock where molten lava flowed into huge depressions caused by giant meteorites, then solidified. Volcanic activity on the moon ceased about two billion years ago.

EARLY MOON MAP
The same side of the Moon always faces towards the Earth. Because the Moon's orbit is not circular and it travels at different speeds, we can see more than half of the Moon. This phenomenon, called libration, means that about 59 per cent of the Moon's surface is visible from the Earth. In 1647 Johannes Hevelius (1611–1687) published his lunar atlas *Selenographia* showing the Moon's librations.

Shadow is used to calculate the height of crater walls

COPERNICUS CRATER
The Moon's craters were formed between 3,500 and 4,500 million years ago by the impact of countless meteorites. These impact craters are all named after famous astronomers and philosophers. Because the Moon has no atmosphere, there has been little erosion of its surface. This plaster model shows Copernicus crater, which is 90 km (56 miles) across and 3,352 m (11,000 ft) deep. Inside the crater there are mountains with peaks 5 km (8 miles) above the crater's floor.

Floor of the crater

Crater walls

Umbra or total shadow

Sun

Moon's orbit

Earth

Moon

Penumbra, or partial shadow

A LUNAR ECLIPSE
An eclipse happens when the Earth passes directly between the Sun and the full Moon, so that the Earth's shadow falls on to the surface of the Moon. This obscures the Moon for the duration of the eclipse.

Lunar equator

Equatorial dial

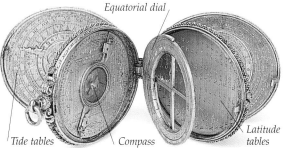

Tide tables   *Compass*   *Latitude tables*

TIDE TABLES
The pull of the Moon's gravity (p.21), and to a lesser extent, the Sun's, causes the water of the seas on Earth to rise and fall. This effect is called a tide. When the Sun, the Moon, and the Earth are all aligned at a New or Full Moon, the tidal "pull" is the greatest. These are called Spring tides. When the Sun and the Moon are at right angles to each other, they produce smaller pulls called Neap tides. This compendium (1569) contains plates with tables indicating the tides of some European cities. It was an essential instrument for sailors entering harbour.

PHASES OF THE MOON

The phases of the Moon are caused by the constantly changing series of angles formed by the Sun and the Moon as the Moon revolves around the Earth. When the Moon and the Sun are on opposite sides of the Earth, the Sun shines directly on the Moon's surface, resulting in a Full Moon. When the area of the lit surface increases, the Moon is said to be waxing; as it decreases, it is said to be waning.

Waxing crescent Moon at 4 days

Full Moon at 14 days

Waning 19-day Moon

Moon at 21 days

Moon at 24 days

Gearing

Meridian circle

THE SURFACE OF THE MOON

The features on the far side of the Moon were a mystery until the late 1950s. This view of the terrain was taken by the *Apollo 11* lunar module in 1969. One of the primary purposes for exploring the Moon was to bring back samples of rock to study them and to discover their origins. The Moon is made up of similar but not identical material to that found on Earth. There is less iron on the Moon, but the major minerals are silicates as they are on Earth (p.43) – though they are slightly different in composition. This discovery supports the most popular theory of the Moon's origin. A small planet, about the size of Mars, is thought to have crashed into the Earth about 4.5 billion years ago. The collision tore debris away from both bodies and the Moon formed from this material.

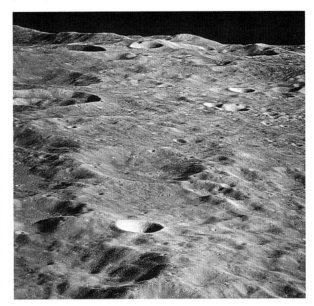

Cross-polarized light in the microscope gives colours

Watery clearness shows no weathering

Hour circle

Earth

INVESTIGATING MOON ROCK

Rocks from the Moon have been investigated by geologists in the same way as they study Earth rocks. The rocks are ground down to thin slices and then looked at under a powerful microscope. The minerals, chiefly feldspar and olivine, which are abundant on the Earth, are unweathered. This is exceptional for geologists because there are no Earth rocks that are totally unweathered.

A MOON GLOBE

Selenography is the study of the surface features of the Moon and this selenograph, created by the artist John Russell in 1797, is a Moon globe. Only a little more than half of the globe is filled with images because at that time the features on the far side of the Moon were unknown. Not until the Russians received the earliest transmissions from the *Luna 3* probe in October 1959 was it possible to see images of what was on the Moon's far side.

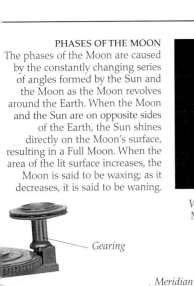

FACTS ABOUT THE MOON

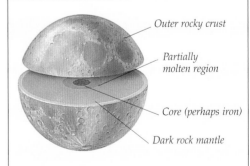

Outer rocky crust

Partially molten region

Core (perhaps iron)

Dark rock mantle

- Interval between two New Moons 29 days 12 hr 44 min
- Temperature at surface −155°C to +105°C
- Rotational period 27.3 Earth days
- Mean distance from the Earth 384,000 km/239,000 miles
- Volume (Earth = 1) 0.02
- Mass (Earth = 1) 0.012
- Density (water = 1) 3.34
- Equatorial diameter 3,476 km/2,160 miles

The Earth

THE EARTH IS THE ONLY PLANET in the Solar System that is capable of supporting advanced life. Its unique combination of liquid water, a rich oxygen- and nitrogen-based atmosphere, and dynamic weather patterns provide the basic elements for a diverse distribution of plant and animal life. Over millions of years landforms and oceans have been constantly changing, mountains have been raised up and eroded away, and continental plates drift across the Earth. The atmosphere acts like a blanket, evening out temperature extremes and keeping warmth in. Without this "greenhouse effect" (p.47), Earth would be about 33°C (60°F) cooler on average. Over the last few decades, scientists have measured a gradual increase in Earth's temperature. Glaciers and polar ice caps have begun to shrink. It is feared that human activity is causing this rapid change by increasing the amount of carbon dioxide and other "greenhouse gases" in the atmosphere.

THE EARTH AND THE MOON
The English astronomer James Bradley (1693–1762) noted that many stars appear to have irregularities in their paths. He deduced that this is due to the effect of observing from an Earth that wobbles on its axis, caused by the gravitational pull of the Moon (p.41).

CONSTANT GEOGRAPHICAL CHANGE
The Earth's crust is made up of a number of plates that are constantly moving because of currents that rise and fall from the molten iron core at the centre of the Earth. Where the plates collide, they can lift the rocky landscape upwards to create mountain ranges that are then eroded into craggy shapes like the Andes in Patagonia. The tensions caused by these movements sometimes result in earthquakes and volcanic activity.

Sahara Desert

Water covers two-thirds of the Earth's surface

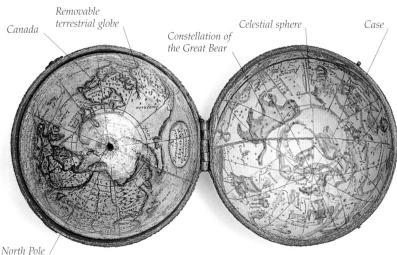

Canada

Removable terrestrial globe

Constellation of the Great Bear

Celestial sphere

Case

North Pole

Cloud layers

POCKET GLOBE
A globe is a convenient tool for recording specific features of the Earth's surface. This 19th-century pocket globe summarizes the face of the world from the geopolitical perspective where the continents are divided into nations and spheres of influence. On the inside of the case is a map of the celestial sphere (pp.12–13), with all the constellations marked out.

Collenia

FOSSILIZED ALGAE
Dead plants and creatures buried in sediment are slowly turned to rock, becoming fossils. This rock contains the fossilized remains of tiny algae that were one of the earliest life forms.

HUMAN DAMAGE
Many scientists wonder if humans, like the dinosaurs, might also become extinct. The dinosaur seems to have been a passive victim of the changing Earth, while humans are playing a key role in the destruction of their environment. In the year 2000 there were more than 6 billion people on the Earth – all producing waste and pollution. In addition to global warming that may be occurring due to the greenhouse effect, chemicals are being released that deplete the ozone layer – a layer in the atmosphere that keeps out dangerous ultraviolet radiation.

EARLY LIFE ON THE EARTH
The first life on the Earth was primitive plants that took carbon dioxide from the air and released oxygen during photosynthesis. Animals evolved when there was enough oxygen in the atmosphere to sustain them. Knowledge about evolving life forms comes in the form of fossils in the rocks (left). However, life forms survive only if environmental conditions on Earth are suitable for them. The dinosaurs, for example, though perfectly adapted to their age became extinct about 65 million years ago.

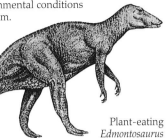

Plant-eating
Edmontosaurus

LIFE-GIVING ATMOSPHERE
Our atmosphere extends out for about 1,000 km (600 miles). It sustains life and protects us from the harmful effects of solar radiation. It has several layers, but the life-sustaining layer is the troposphere, up to 10 km (6 miles) above the Earth's surface.

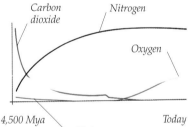

Carbon dioxide Nitrogen

Oxygen

4,500 Mya   Today

Hydrogen

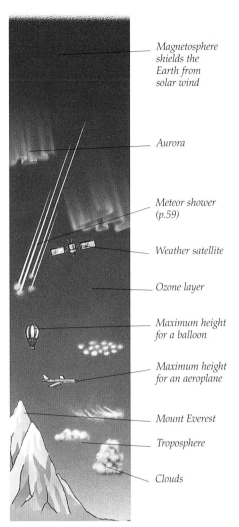

Magnetosphere shields the Earth from solar wind

Aurora

Meteor shower (p.59)

Weather satellite

Ozone layer

Maximum height for a balloon

Maximum height for an aeroplane

Mount Everest

Troposphere

Clouds

EVOLUTION OF THE ATMOSPHERE
Since the Earth was formed, the chemical make-up of the atmosphere has evolved. Carbon dioxide (CO_2) decreased significantly between 4,500 and 3,000 million years ago (Mya). There was a comparable rise in nitrogen. The levels of oxygen began to rise at the same time, due to photosynthesis of primitive plants which used up CO_2 and released oxygen.

THE SPHERICAL EARTH
As early as the 5th century BCE the Greek philosophers had proposed that the Earth is spherical and by the 3rd century BCE they had worked out a series of experiments to prove it. But it was not until the first satellites were launched in the late 1950s that humans saw what their planet looks like from space. The one feature that makes the Earth unique is the great abundance of liquid water; more than two-thirds of the surface is covered with water. Water makes the Earth a dynamic place. Erosion, tides, weather patterns, and plentiful forms of life are all tied to the presence of water. There is more water in the Sahara Desert in North Africa than there is on Venus (pp.46–47).

FACTS ABOUT THE EARTH

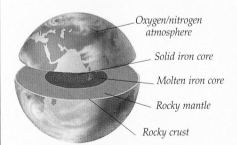

Oxygen/nitrogen atmosphere

Solid iron core

Molten iron core

Rocky mantle

Rocky crust

- **Sidereal period** 365.26 days
- **Temperature** –70°C to +55°C
- **Rotational period** 23 hr 56 min
- **Mean distance from the Sun** 149.6 million km/93 million miles
- **Volume** 1 • **Mass** 1
- **Density** (water = 1) 5.52
- **Equatorial diameter** 12,760 km/7,930 miles
- **Number of satellites** 1 (the moon)

Mercury

THE PLANET MERCURY IS NAMED after the Greco-Roman messenger of the gods, because it circles the Sun faster than the other planets, completing its circuit in 88 Earth days. Because it travels so close to the Sun, Mercury is often difficult to observe. Even though its reflected light makes it one of the brightest objects in the night sky, Mercury is never far enough from the Sun to be able to shine out brightly. It is only visible as a "morning" or "evening" star, hugging the horizon just before or after the Sun rises or sets. Like Venus, Mercury also has phases (p.20). Being so close to the Sun, temperatures during the day on Mercury are hot enough to melt many metals. At night they drop to –180°C (–291°F) making the temperature range the greatest of all the planets. The gravitational pull of the Sun has "stolen" any atmosphere that Mercury had to protect itself against these extremes.

EARLY MERCURY MAP
Although many astronomers have tried to record the elusive face of Mercury, the most prolific observer was the French astronomer Eugène Antoniadi (1870–1944). His maps, drawn between 1924 and 1929, show a number of huge valleys and deserts. Close-up views by the *Mariner 10* space probe uncovered an altogether different picture (below).

SEISMIC WAVES
Some of Mercury's hills and mountains were created by the impact of a huge meteorite (p.59). The impact created a crater, known as Caloris Basin, where the meteorite struck the surface and sent out seismic, or shock, waves through the semi-molten core of the planet. These waves travelled through Mercury to the other side, where the crust buckled and mountain ranges were thrown up.

Caloris impact

Mountain range

Seismic waves

CRATERED TERRAIN
The surface of Mercury closely resembles our crater-covered Moon (p.40). Mercury's craters were also formed by the impact of meteorites, and the lack of atmosphere has kept the landscape unchanged. Around the edges of the craters, a series of concentric ridges record how the surface was pushed outwards by the force of the impact.

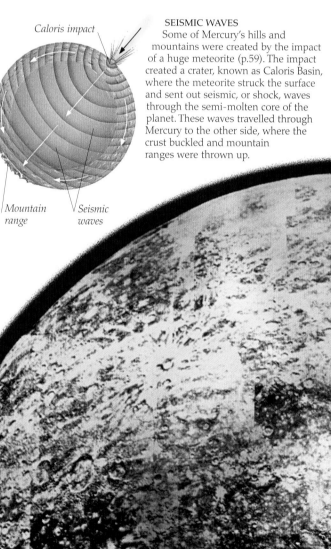

SURFACE OF MERCURY
This image is a mosaic of photographs taken during *Mariner 10*'s journey past Mercury in 1974. Mercury seems to have shrunk a great deal after it was formed. This has caused a series of winding ridges, called scarps, that are unique to the planet. The entire surface is heavily cratered. The space probe *Mariner 10* also discovered that Mercury has a magnetic field about 1 per cent the strength of the Earth's magnetic field.

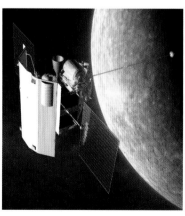

MESSENGER TO MERCURY
In 2004, NASA launched the *Messenger* spacecraft to explore Mercury. After flying past Mercury three times it will go into orbit around the planet in 2011. It is the first mission to Mercury since *Mariner 10* in 1974–75.

LOOKING AT VOLUME

These blocks – wood, aluminium and iron – all have the same volume, that is they occupy the same amount of space. Despite being the same size, however, these materials do not have the same mass and density, nor do they weigh the same. This is also true of the planets. For example, Mercury, though small, has a higher density than that of some of the larger planets.

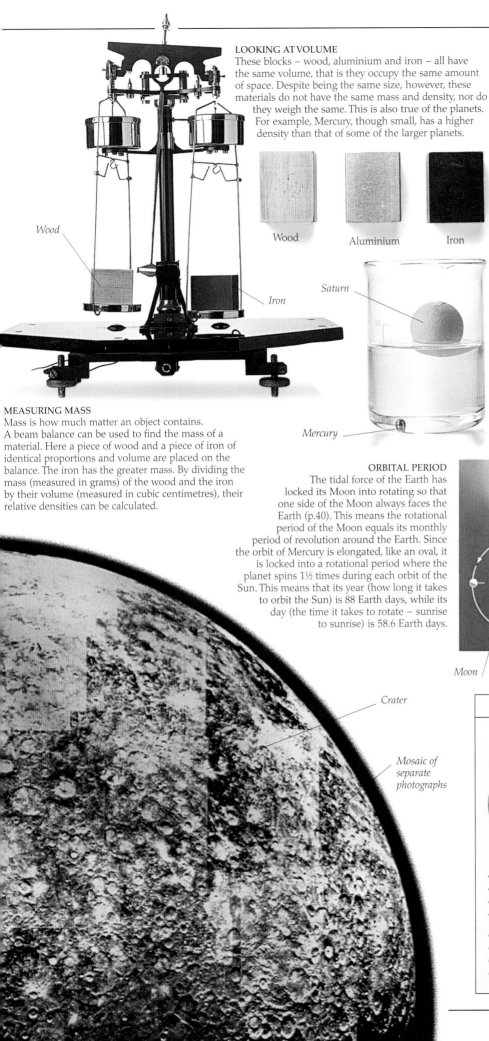

Wood

Iron

Wood

Aluminium

Iron

Measuring planets

Whereas we can weigh and measure objects on the Earth, we have to assess the space a planet occupies (volume), how much matter it contains (mass), and its density by looking at its behaviour, by analyzing its gravitational pull on nearby objects, and by using data gained by space probes (pp.34–35). Density is the mass for every unit of volume of an object (mass divided by the volume).

Saturn

Mercury

COMPARING DENSITY

Mercury has great mass for its size. Even though it is only slightly larger than the Earth's Moon, its mass is four times that of the Moon. This means its density must be nearly as high as the Earth's, most likely due to a very high quantity of iron. Astronomers believe that Mercury must have a massive iron core that takes up nearly three quarters of its radius to achieve such great mass – a fact backed up by *Mariner 10*'s evidence of a magnetic field. When the densities of Mercury and Saturn, the huge gas giant (pp.52–53), are compared, Saturn would float and Mercury, whose density is seven times as great, would sink.

MEASURING MASS

Mass is how much matter an object contains. A beam balance can be used to find the mass of a material. Here a piece of wood and a piece of iron of identical proportions and volume are placed on the balance. The iron has the greater mass. By dividing the mass (measured in grams) of the wood and the iron by their volume (measured in cubic centimetres), their relative densities can be calculated.

ORBITAL PERIOD

The tidal force of the Earth has locked its Moon into rotating so that one side of the Moon always faces the Earth (p.40). This means the rotational period of the Moon equals its monthly period of revolution around the Earth. Since the orbit of Mercury is elongated, like an oval, it is locked into a rotational period where the planet spins 1½ times during each orbit of the Sun. This means that its year (how long it takes to orbit the Sun) is 88 Earth days, while its day (the time it takes to rotate – sunrise to sunrise) is 58.6 Earth days.

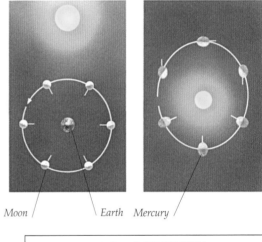

Moon   *Earth*   *Mercury*

Crater

Mosaic of separate photographs

FACTS ABOUT MERCURY

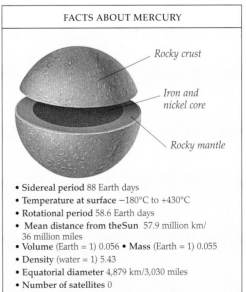

Rocky crust

Iron and nickel core

Rocky mantle

- **Sidereal period** 88 Earth days
- **Temperature at surface** −180°C to +430°C
- **Rotational period** 58.6 Earth days
- **Mean distance from the Sun** 57.9 million km/ 36 million miles
- **Volume** (Earth = 1) 0.056 • **Mass** (Earth = 1) 0.055
- **Density** (water = 1) 5.43
- **Equatorial diameter** 4,879 km/3,030 miles
- **Number of satellites** 0

Venus

People often mistake Venus for a star. After the Moon, it is the brightest object in our night sky. Because it is so close in size to the Earth, until the 20th century astronomers assumed that it might be in some ways like the Earth. The probes sent to investigate have shown that this is not so. The dense cloudy atmosphere of Venus hides its surface from even the most powerful telescope. Only radar can penetrate to map the planet's features. Until it became possible to determine the surface features – largely flat, volcanic plains – scientists could not tell how long the Venusian day was. The atmosphere is deadly. It is made up of a mixture of carbon dioxide and sulphuric acid which causes an extreme "greenhouse effect" in which heat is trapped by the lethal atmosphere. The ancients, however, saw only a beautifully bright planet and so they named it after their goddess of love. Nearly all the features mapped on the surface of Venus have been named after women, such as Pavlova, Sappho, and Phoebe.

VENUS IN THE NIGHT SKY
This photograph was taken from the Earth. It shows the crescent Moon with Venus in the upper left of the sky. Shining like a lantern at twilight, Venus looks so attractive that astronomers were inspired to believe it must be a beautiful planet.

CALCULATING DISTANCES
One way to calculate the distance of the Earth from the Sun is for a number of observers all around the world to measure the transit of a planet (the passage of the planet as it crosses the disc of the Sun and appears in silhouette). The British explorer Captain James Cook led one of the many expeditions in 1769 to observe the transit of Venus from Tahiti. Calculations made from these observations also enabled astronomers to work out the relative measurements of the entire Solar System.

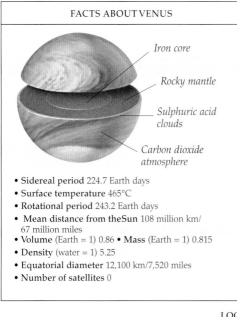

FACTS ABOUT VENUS

- *Iron core*
- *Rocky mantle*
- *Sulphuric acid clouds*
- *Carbon dioxide atmosphere*

- **Sidereal period** 224.7 Earth days
- **Surface temperature** 465°C
- **Rotational period** 243.2 Earth days
- **Mean distance from theSun** 108 million km/ 67 million miles
- **Volume** (Earth = 1) 0.86 • **Mass** (Earth = 1) 0.815
- **Density** (water = 1) 5.25
- **Equatorial diameter** 12,100 km/7,520 miles
- **Number of satellites** 0

Blue-filtered colour

Dense clouds

LOOKING AT VENUS
In 1978 the United States launched the *Pioneer* Orbiter, designed to map the surface of Venus by using radar to penetrate its densely clouded atmosphere. It was followed in 1989 by *Magellan*, which circled Venus every 3 hours and 9 minutes and had a 3.7-m (12-ft) radar dish that beamed radar images back to the Earth for analysis. Computers were used to build up pictures of the surface – mainly volcanic plains. This view from space does not show the true colour of the planet; a blue filter has been used to emphasize the cloud layers. Another Venus mapper is the large radio telescope near Arecibo in Puerto Rico (p.32).

PUZZLING SURFACE

Even with the best telescope, Venus looks almost blank. This led the Russian astronomer, Mikhail Lomonosov (p.28) to propose that the Venusian surface is densely covered with cloud. As recently as 1955 the British astronomer Fred Hoyle (1915–2001) argued that the clouds are actually drops of oil and that Venus has oceans of oil. In fact, the clouds are droplets of dilute sulphuric acid, and the planet has a hot, dry volcanic surface.

ASSEMBLING VENERA PROBES

During the 1960s and 1970s the former USSR sent a number of probes called *Venera* to investigate the surface of Venus. They were surprised when three of the probes stopped functioning as soon as they entered the Venusian atmosphere. Later *Venera* probes showed the reason why – the atmospheric pressure on the planet was 90 times that of Earth, the atmosphere itself was heavily acidic, and the temperature was 465°C (900°F).

Carbon dioxide atmosphere lets heat radiation in but not out

Sunlight reflected

Sulphuric acid layer

Feet of probe

Colour balance

GREENHOUSE EFFECT

The great amount of carbon dioxide in Venus's atmosphere means that, while solar energy can penetrate, heat cannot escape. This has led to a runaway "greenhouse effect". Temperatures on the surface easily reach 465°C (900°F), even though the thick cloud layers keep out as much as 80 per cent of the Sun's rays.

Hot surface

Infrared radiation

Sif Mons volcano

LANDING ON VENUS

This image was sent back by *Venera 13* when it landed on Venus in 1982. Part of the space probe can be seen bottom left and the colour balance, or scale, is in the lower middle of the picture. The landscape appears barren, made up of volcanic rocks. There was plenty of light for photography, but the spacecraft succumbed to the oven-like conditions after only an hour.

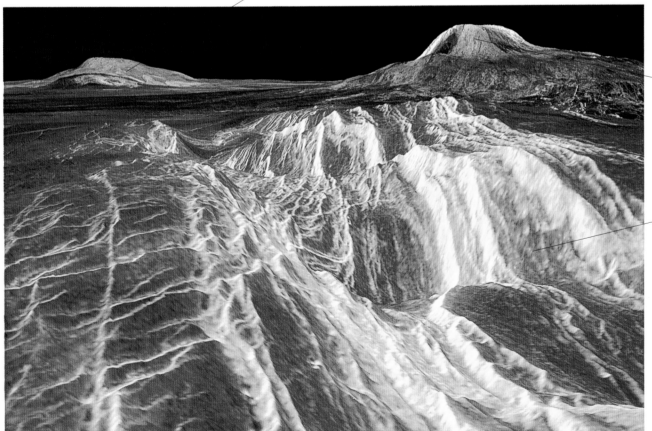

Gula Mons volcano

Lava flows

THREE-DIMENSIONAL VIEW

This radar image of the Western Eistla Region, sent by *Magellan*, shows the volcanic lava flows (see here as the bright features) that cover the landscape and blanket the original Venusian features. Most of the landscape is covered by shallow craters. The simulated colours are based on those recorded by the Soviet *Venera* probes.

Mars

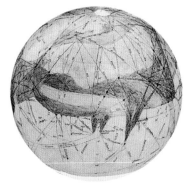

Mᴀʀs ᴀᴘᴘᴇᴀʀs ᴘᴀʟᴇ ᴏʀᴀɴɢᴇ ɪɴ ᴛʜᴇ ɴɪɢʜᴛ sᴋʏ. The Babylonians, Greeks, and Romans all named it after their gods of war. In reality, Mars is a small planet – only half the size of the Earth – but there are similarities. Mars, like the Earth, has a 24-hour day, polar caps, and an atmosphere. Not surprisingly, Mars has always been the most popular candidate as a site for possible extraterrestrial life. Many scientists believe that some form of life - or at least evidence of past life – may remain within the planet, but no life could survive on the surface. The atmosphere is too thin to block out deadly ultraviolet rays. Mars is also further from the Sun than the Earth, making it a lot colder.

Arabia region

MARTIAN MARKINGS
In 1659 the Dutch scientist Christiaan Huygens (1629–1695) drew the first map of Mars, showing a V-shaped mark on the surface that reappeared in the same place every 24 hours. This was Syrtis Major. He concluded, correctly, that its regular appearance indicated the length of the Martian day. The American astronomer, Percival Lowell (1855–1916), made a beautiful series of drawings of the Martian canals (above) described by Schiaparelli (see below). Closer inspections showed that these canals were optical illusions caused by the eye joining up non-related spots.

CANALS ON MARS
The Italian astronomer Giovanni Schiaparelli (1835–1910) made a close study of the surface of Mars. In 1877 he noticed a series of dark lines that seemed to form some sort of network. Schiaparelli called them *canali*, translated as "channels" or "canals". This optical illusion seems to be the origin of the myth that Mars is occupied by a sophisticated race of hydraulic engineers. It was Eugène Antoniadi (p.44) who made the first accurate map of Mars.

Water ice

Ice cap

Cliff

AROUND THE PLANET MARS
The Martian atmosphere is much thinner than that of the Earth and is composed mostly of carbon dioxide. There is enough water vapour for occasional mist, fog, and cloud to form. *Mariner 9*, the first spacecraft to orbit Mars, revealed a series of winding valleys in the Chryse region that could be dried-up river beds. Mars also has large volcanoes. One of them – Olympus Mons – is the largest in the Solar System. There are also deserts, canyons, and polar ice caps.

POLAR ICE
The polar regions of Mars are covered by a thin layer of ice, which is a mixture of frozen water and solid carbon dioxide. This image of the north polar ice cap, taken by ESA's *Mars Express* spacecraft, shows layers of water ice, dune fields and cliffs almost 2 km (1 mile) high. The polar caps are not constant, but grow and recede with the Martian seasons.

Computer-processed view of Mars from *Viking* orbiters (1980)

Robotic arm

SAMPLING ROCK
In 1997, *Pathfinder* landed on Mars with a 63-cm (25-in) long robot rover called *Sojourner*. The rover carried special instruments so that it could analyse the composition of Martian rocks.

TESTING FOR LIFE
The two *Viking* probes in the 1970s carried out simple experiments on Martian soil. They found no signs of life.

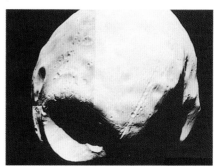

GLOBAL SURVEYOR
Mars Global Surveyor returned thousands of high-resolution images of Mars between 1999 and 2006. It also studied the planet's weather and chemical make-up.

Assembling the *Viking* lander

DESERT LANDSCAPE
Mars resembles a desert. Winds whip up the red dust and it becomes suspended in the atmosphere, giving the sky a reddish hue. To make sure that the *Viking* images could be reproduced in their proper colours, the spacecraft carried a series of colour patches (p.47). Its photographs of these were corrected until the patches were in their known colours, so the scientists could be confident that the landscape colours were also shown correctly.

MARTIAN MOONS
Mars has two small moons, Phobos (right) and Deimos, 28 and 16 km (17 and 10 miles) in diameter. Since the orbit of Deimos is only 23,460 km (14,580 miles) from the centre of Mars, it will probably be pulled down to the surface with a crash in about 50 million years' time.

GULLIES ON MARS
Images sent back by *Mars Global Surveyor* show these intriguing marks. They are gullies on the wall of a meteor impact crater. It is possible that they formed when the permafrost beneath the surface melted, allowing ground water up to the surface. They provided evidence for the existence of water on Mars. The ripples at the bottom of the picture are sand dunes.

FACTS ABOUT MARS

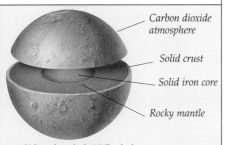

Carbon dioxide atmosphere

Solid crust

Solid iron core

Rocky mantle

- **Sidereal period** 687 Earth days
- **Surface temperature** −120°C to +25°C
- **Rotational period** 24 hr 37 min
- **Mean distance from the Sun** 230 million km/ 141 million miles
- **Volume** (Earth=1) 0.15
- **Mass** (Earth=1) 0.11
- **Density** (water=1) 3.95
- **Equatorial diameter** 6,790 km/4,220 miles
- **Number of satellites** 2

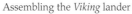

Jupiter

THIS HUGE, BRIGHT PLANET is the largest world in our Solar System; four of its moons are the size of planets. It is different in structure from the solid inner planets. Apart from a small rocky core, Jupiter is mainly hydrogen and helium. Below the cloudy atmosphere, the pressure is so great that these are liquid rather than gas. Deep down, the liquid hydrogen behaves like a metal. As a result, Jupiter has a strong magnetic field and fierce radiation belts. Jupiter emits more heat radiation than it receives from the Sun, because it continues shrinking at a rate of a few millimetres a year. Had Jupiter been only 13 times more massive, this contraction would have made the centre hot enough for nuclear fusion reactions (p.38) to begin, though not to be sustained for as long as in a star. It would have become a brown dwarf – a body between a planet and a star. The *Galileo* spacecraft, which orbited Jupiter from 1995–2003, transmitted some amazing photographs of Jupiter and its moons.

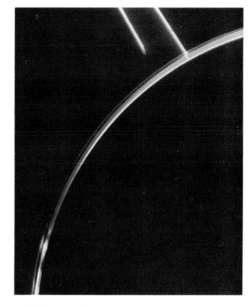

JUPITER'S RINGS
The US *Pioneer* missions were sent past Jupiter in the early 1970s, *Pioneer 10* sending back the first pictures. In 1977 the US sent two *Voyager* probes to explore Jupiter's cloud tops and five of its moons. *Voyager 1* uncovered a faint ring – like Saturn's rings (p.53) – circling the planet. The thin yellow ring (approximately 30 km/18 miles thick) can be seen at the top of the photograph.

SEEING THE RED SPOT
In 1660 the English scientist, Robert Hooke (1635–1702), reported seeing "a spot in the largest of the three belts of Jupiter". Gian Cassini (p.28) saw the spot at the same time, but subsequent astronomers were unable to find it. The Great Red Spot was observed again in 1878 by the American astronomer Edward Barnard (1857–1923).

FACTS ABOUT JUPITER

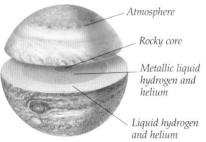

- Atmosphere
- Rocky core
- Metallic liquid hydrogen and helium
- Liquid hydrogen and helium

- **Sidereal period** 11.86 Earth years
- **Temperature at cloud tops** –150°C
- **Rotational period** 9 hr 55 min
- **Mean distance from the Sun** 778 million km/ 484 million miles
- **Volume** (Earth = 1) 1,319 • **Mass** (Earth = 1) 318
- **Density** (water = 1) 1.33
- **Equatorial diameter** 142,980 km/89,350 miles
- **Number of satellites** at least 63

North Polar region
North Temperate Belt
North Tropical Zone
Storm system
Equatorial Belt
Equatorial Zone
South Tropical Zone
South Temperate Belt
Great Red Spot
South Polar region

JUPITER'S CLOUDS
The cloud tops of Jupiter seem to be divided into a series of bands that are different colours. The light bands are called zones, and the dark bands belts. The North Tropical Zone (equivalent to our northern temperate zone) is the brightest, its whiteness indicating high-level ammonia clouds. The Equatorial Belt, surrounding Jupiter's equator, always seems in turmoil, with the atmosphere constantly whipped up by violent winds. Across the planet are a number of white or red ovals. These are huge cloud systems. The brown and orange bands indicate the presence of organic molecules including ethane.

An impact site

A MOMENTOUS IMPACT

In July 1994, fragments of the comet Shoemaker-Levy 9 crashed into Jupiter's southern hemisphere at speeds of around 210,000 km/h (130,500 mph). The comet had been discovered in 1993 by astronomers Carolyn and Eugene Shoemaker and David Levy, who also predicted its path. It was the first time in history that astronomers had been able to predict a collision between two bodies in the Solar System and then observe the event. Over 20 pieces of the comet hit Jupiter, some of them sending up 3,000-km (1,865-mile) high fireballs and plumes.

Jupiter's moons

In 1610 Galileo (p.20) made the first systematic study of the four largest moons of Jupiter. Since they seemed to change their positions relative to the planet every night, he concluded, correctly, that these objects must be revolving around Jupiter. This insight provided more ammunition for the dismantling of the geocentric theory (p.11), which placed the Earth at the centre of the Universe. In 1892 another small moon was discovered circling close to the cloud tops of the planet. To date, a total of 63 moons have been discovered.

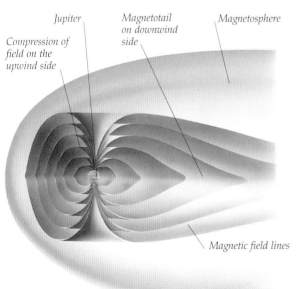

Jupiter   *Magnetotail on downwind side*   *Magnetosphere*

Compression of field on the upwind side

Magnetic field lines

SPINNING JUPITER

Jupiter spins so quickly that its day is only 9 hours and 55 minutes long and its equator bulges outwards. Another effect of the rapid rotation is that the spinning of Jupiter's metallic hydrogen core generates a huge magnetic field around the planet. This magnetosphere is pushed back by the solar wind and its tail spreads out over a vast distance, away from the Sun.

CALLISTO

Callisto is the second largest of Jupiter's moons, and the most heavily cratered, not unlike our Moon, except that the craters are made of ice. The bright areas are the ice craters formed by impacts of objects from space.

ERUPTION ON IO

Io is the Moon that is closest to Jupiter. It is one of the "Galilean" moons, named after Galileo, who discovered them. The others are Callisto, Europa, and Ganymede. The erupting plume of a massive volcano can be seen here on the horizon, throwing sulphuric materials 300 km (186 miles) out into space. The photograph was taken by *Voyager* from a distance of half a million kilometres (310,700 miles) and has been specially coloured using filters.

Saturn

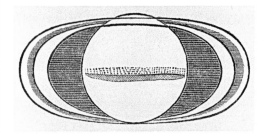

17TH-CENTURY VIEW
In 1675 the Bolognese director of the Paris Observatory, Gian Domenico Cassini (p.28), discovered that, despite appearances, Saturn did not have a single, solid ring. He could see two rings, with a dark gap in between. His drawing, made in 1676, shows the gap, which was called the Cassini Division in his honour.

THE GIANT PLANET SATURN, with its flat rings, is probably the most widely recognized astronomical image. For the classical world, Saturn was the most distant planet known. They named it after the original father of all the gods. Early astronomers noted its 29-year orbit and assumed that it moves sluggishly. Composed mostly of hydrogen, its atmosphere and structure are similar to Jupiter's, but its density is much lower. Saturn is so light that it could float on water (p.45). Like Jupiter, Saturn rotates at great speed causing its equator to bulge outwards. Saturn also has an appreciable magnetic field. Winds in its upper atmosphere can travel at 1,800 km/h (1,100 mph) but major storms are rare. White spots tend to develop during Saturn's northern-hemisphere summer, which happens every 30 years or so, the last being in 1990.

SATURN AND THE RINGS
Though Saturn's rings look solid from Earth, astronomers have known since the 19th century that they cannot be. In fact, they consist of countless individual particles, made of ice and dust, ranging in size from specks to hundreds of metres. The rings are only about 30 m (100 ft) thick, but their total width is more than 272,000 km (169,000 miles).

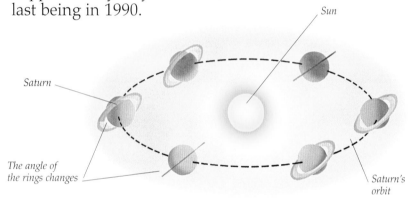

Sun

Saturn

The angle of the rings changes

Saturn's orbit

CHANGING VIEW
Saturn's axis is tilted. Because the rings lie around its equator, they incline as the planet tilts. This means that the rings change dramatically in appearance, depending on what time during Saturn's year they are being observed – Saturn's year is equal to 29.4 Earth years. The angle of the rings appears to change according to how Saturn and the Earth are placed in their respective orbits.

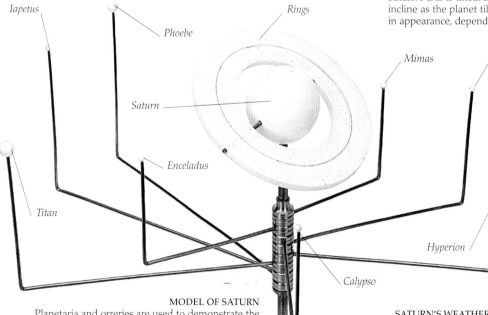

Iapetus

Phoebe

Rings

Mimas

Tethys

Saturn

Enceladus

Titan

Hyperion

Calypso

MODEL OF SATURN
Planetaria and orreries are used to demonstrate the shapes and satellites of the planets. This orrery shows Saturn with the eight moons that were known in the 19th century. The model is flawed, however, because it is impossible to show the relative size of a planet and the orbits of its satellites.

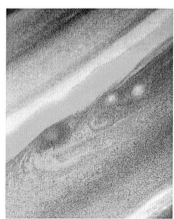

SATURN'S WEATHER PATTERNS
The weather patterns of Saturn's northern hemisphere were photographed from a range of 7 million km (4.4 million miles) by *Voyager 2* in 1981. Storm clouds and white spots are visible features of Saturn's weather.

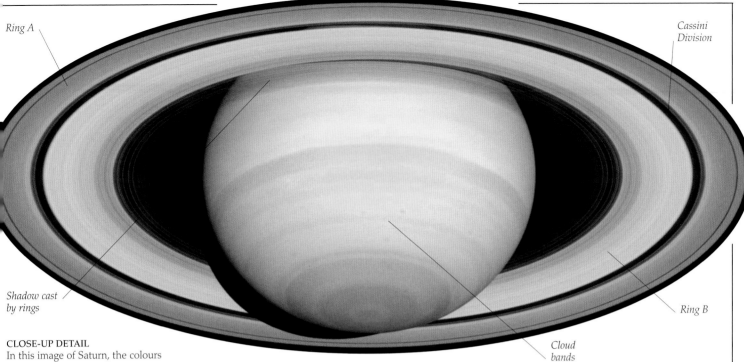

Ring A

Cassini Division

Shadow cast by rings

Cloud bands

Ring B

CLOSE-UP DETAIL

In this image of Saturn, the colours are exaggerated to show more clearly the cloud bands encircling the planet. We can also see that the rings are made up of many separate ringlets. The principal rings, called A and B, are easily visible from Earth with a small telescope. Saturn also has five fainter rings. As Saturn orbits the Sun, we see the ring system from different angles. Sometimes the rings look open, as in this image. Every 15 years, the rings are presented to us edge-on and they virtually disappear from view. The most detailed images of Saturn have been returned by the *Cassini-Huygens* mission, which was launched in 1997 and arrived in 2004. The *Huygens* probe was released, and landed on Titan, Saturn's largest moon, which is hidden by opaque haze. Using radar, the orbiting *Cassini* spacecraft has discovered that Titan has lakes of liquid methane.

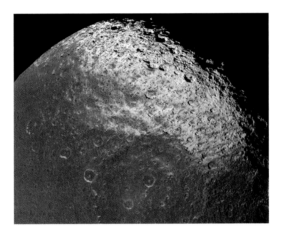

TWO-TONE MOON

Iapetus is Saturn's third largest Moon, with a diameter of 1,436 km (892 miles). It is made mostly of ice. One of its strangest features is that one half of the surface is very much darker that the other. The dark area is coated with material as black as tar, which seems to have fallen on it. This picture was taken by the *Cassini* spacecraft. *Cassini* revealed a range of mountains up to 20 km (12 miles) high extending for about 1,300 km (800 miles).

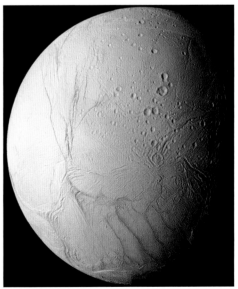

FACTS ABOUT SATURN

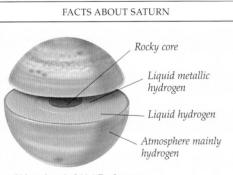

Rocky core

Liquid metallic hydrogen

Liquid hydrogen

Atmosphere mainly hydrogen

- **Sidereal period** 29.4 Earth years
- **Temperature at cloud tops** −180°C
- **Rotational period** 10 hr 40 min
- **Mean distance from the Sun** 1,430 million km/ 886 million miles
- **Volume** (Earth = 1) 744 • **Mass** (Earth = 1) 95.18
- **Density** (water = 1) 0.69
- **Equatorial diameter** 120,535 km/74,900 miles
- **Number of satellites** 60

SEEING THE RINGS

When Galileo first discovered Saturn's rings in 1610, he misinterpreted what he saw. He thought Saturn was a triple planet. It was not until 1655 that the rings were successfully identified and described by the Dutch scientist and astronomer Christiaan Huygens (1629–1695), using a powerful telescope that he built himself.

TIGER STRIPES

Saturn's moon, Enceladus, is about 500 km (310 miles) across. This false colour image of its icy surface from the *Cassini* spacecraft reveals a series of parallel fissures (in blue), which astronomers nicknamed "tiger stripes". Other images have shown plumes of icy droplets jetting out of these fissures from liquid water below the frozen crust. Large areas of the surface have no craters, or very few. This means that it has greatly altered since Enceladus first formed.

Uranus

URANUS WAS THE FIRST PLANET to be discovered since the use of the telescope. It was discovered by accident, when William Herschel, observing from Bath, England, set about remeasuring all the major stars with his 15-cm (6-in) reflector telescope (p.24). In 1781 he noticed an unusually bright object in the zodiacal constellation of Gemini. At first he assumed it was a nebula (pp.60–61) and then a comet (pp.58–59), but it moved in a peculiar way. The name of Uranus was suggested by the German astronomer, Johann Bode, who proposed that the planet be named after the father of Saturn, in line with established classical traditions. Bode is also famous as the creator of Bode's Law – a mathematical formula that predicted roughly where planets should lie.

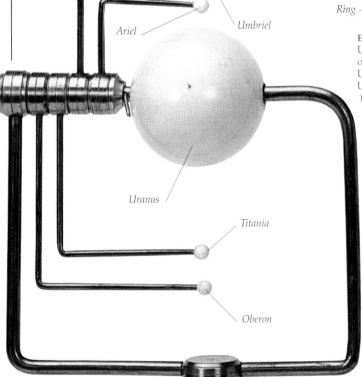

WILLIAM HERSCHEL (1738–1822)
William Herschel was so impressed by a treatise on optics, which described the construction of telescopes, that he wanted to buy his own telescope. He found them too expensive, so in 1773 he decided to start building his own. From that moment on, astronomy became Herschel's passion.

Ariel

Umbriel

Uranus

Titania

Oberon

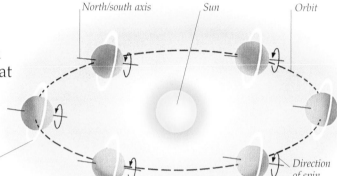

North/south axis   *Sun*   *Orbit*

Ring   *Direction of spin*

ECCENTRIC TILT
Uranus spins on an axis that is tilted at an angle of nearly 98° from the plane of its orbit. This means that, compared with all the other planets in the Solar System, Uranus is spinning on its side. During its 84-year orbit of the Sun, the North Pole of Uranus will have 42 years of continuous, sunny summer, while the South Pole has the same length of sunless winter, before they swap seasons. This odd tilt may be the result of a catastrophic collision during the formation of the Solar System.

19TH-CENTURY MODEL
Because of the odd angle of Uranus's rotational axis, all its known satellites also revolve at right angles to this axis, around Uranus's equator. This fact is demonstrated by an early model, which shows the planet and four of its moons tilted at 98°. This orrery (p.36) dates from the 19th century when only four of the 27 moons had been discovered.

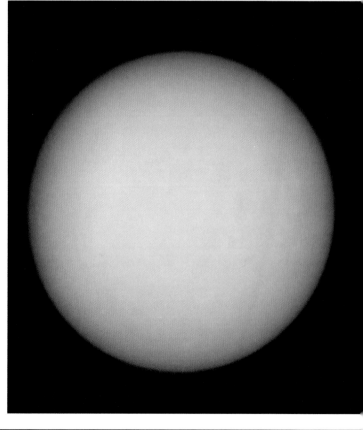

VIEW FROM SPACE
Uranus is a giant planet, four times larger than Earth. The Hubble Space Telescope took these contrasting views in 2004. On the left, Uranus is seen in natural colour. It looks blue because of absorption by methane in the atmosphere. The image on the right is false colour, which shows up bright clouds and hazy bands parallel to the equator.

AIRBORNE OBSERVATION OF URANUS
The covering of one celestial body by another is known as occultation. A team of scientists observed the occultation of a star by Uranus in 1977 from NASA's Kuiper Airborne Observatory over the Indian Ocean. This was when the faint rings of Uranus were observed for the first time.

LITERARY MOONS
All the satellites of Uranus are named after sprites and spirits drawn from English literature. The American astronomer Gerard P. Kuiper (1905–1973) discovered Miranda in 1948. (Miranda and Ariel are characters from William Shakespeare's *The Tempest*.) It has a landscape unlike any other in the Solar System. Miranda seems to be composed of a jumble of large blocks. Scientists have suggested that these were caused by some huge impact during which Miranda was literally blown apart. The pieces drifted back together through gravitational attraction, forming this strange mixture of rock and ice.

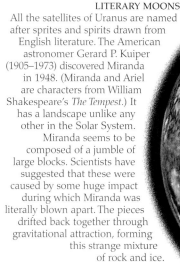

URANUS RING SYSTEM
While watching the occultation of Uranus in 1977, astronomers noticed that the faint star "blinked on and off" several times at the beginning and end of the occultation. They concluded that Uranus must have a series of faint rings that caused the star to "blink" by blocking off its light as it passed behind them. The *Voyager 2* flyby in 1986 uncovered two more rings. The rings of Uranus are thin and dark, made up of particles only about 1 metre (39 in) across. The broad bands of dust between each ring suggest that the rings are slowly eroding.

THE MOON TITANIA
William Herschel discovered Uranus's two largest moons in 1789, naming them Oberon and Titania, the fairy king and queen in William Shakespeare's *A Midsummer Night's Dream*. The English astronomer William Lassell (1799-1880) discovered Ariel and Umbriel in 1851. Miranda was discovered in 1948. Since then, a further 22 moons have been found.

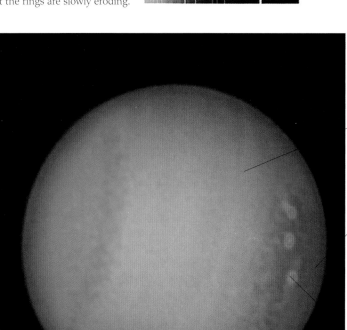

Hazy cloud bands

Polar region

Bright clouds

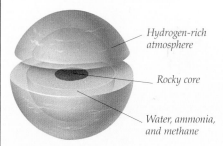

FACTS ABOUT URANUS

Hydrogen-rich atmosphere

Rocky core

Water, ammonia, and methane

- **Sidereal period** 83.8 Earth years
- **Temperature at cloud tops** – 210°C
- **Rotational period** 17 hr 14 min
- **Mean distance from the Sun** 2,870 million km/ 1,785 million miles
- **Volume** (Earth = 1) 67
- **Mass** (Earth =1) 14.5
- **Density** (water = 1) 1.29
- **Equatorial diameter** 51,120 km/31,765 miles
- **Number of satellites** 27

Neptune and beyond

NEPTUNE WAS DISCOVERED AS THE RESULT of calculations. By the early 19th century, astronomers realized that Uranus was not following its expected orbit. The gravitational pull of an unknown planet beyond Uranus seemed the most likely explanation. In 1845, the English mathematician John Couch Adams (1819–1892) announced that he had calculated the probable position of a planet beyond Neptune but his findings were ignored. In June 1846, the Frenchman Urbain Le Verrier did the same. This time, observers took notice. Johann Galle (1812–1910) of the Berlin Observatory found Neptune on 23 September 1846. Astronomers continued to speculate about another planet beyond Neptune. Pluto was eventually discovered in 1930 and was considered to be the ninth major planet until 2006. Between 1992 and 2006, hundreds of small icy bodies had been found beyond Neptune, in what is called the Kuiper belt. They include Eris, which is larger than Pluto. In 2006, astronomers decided to class both Pluto and Eris as dwarf planets.

URBAIN LE VERRIER (1811–1877)
Le Verrier was a teacher in chemistry and astronomy at the *Ecole Polytechnique*. Having calculated the position of Neptune, Le Verrier relied on others to do the actual "looking" for the planet for him.

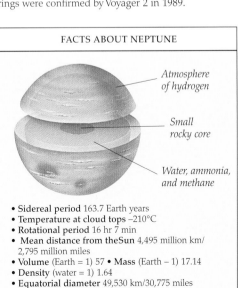

NEPTUNE'S RINGS
Neptune, like all the giant planets, has a series of rings encircling it. The rings were discovered when the planet passed in front of a star. Results of an occultation (p.55) in July 1984 showed the typical "blinking on and off", indicating that Neptune's rings were blocking out the light of the distant star. There seem to be two main rings, with two faint inner rings. The inner ring is less than 15 km (9 miles) wide. The rings were confirmed by Voyager 2 in 1989.

FACTS ABOUT NEPTUNE

- Atmosphere of hydrogen
- Small rocky core
- Water, ammonia, and methane

- **Sidereal period** 163.7 Earth years
- **Temperature at cloud tops** –210°C
- **Rotational period** 16 hr 7 min
- **Mean distance from theSun** 4,495 million km/ 2,795 million miles
- **Volume** (Earth = 1) 57 • **Mass** (Earth – 1) 17.14
- **Density** (water = 1) 1.64
- **Equatorial diameter** 49,530 km/30,775 miles
- **Number of satellites** 13

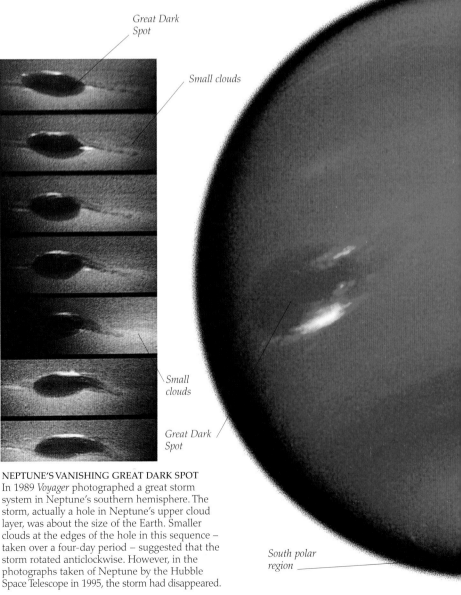

Great Dark Spot

Small clouds

Small clouds

Great Dark Spot

South polar region

NEPTUNE'S VANISHING GREAT DARK SPOT
In 1989 *Voyager* photographed a great storm system in Neptune's southern hemisphere. The storm, actually a hole in Neptune's upper cloud layer, was about the size of the Earth. Smaller clouds at the edges of the hole in this sequence – taken over a four-day period – suggested that the storm rotated anticlockwise. However, in the photographs taken of Neptune by the Hubble Space Telescope in 1995, the storm had disappeared.

Discovering Pluto

Pluto was discovered in 1930 as the result of a systematic search by an American astronomer, Clyde Tombaugh (1906–1997), working at the Lowell Observatory in Arizona. Its orbit was found to be unusual, being much more elongated than the orbits of the previously known planets. Pluto is sometimes nearer to the Sun than Neptune. Gradually, astronomers realised that Pluto was much smaller than they originally thought. It has only one fifth the mass of our Moon. The first spacecraft ever to be sent to Pluto, *New Horizons*, was launched in 2006 and will fly by Pluto in 2015.

THE KUIPER BELT
In 1951, the Dutch-American astronomer Gerard Kuiper (1905–1973) predicted the existence of a whole belt of small icy worlds beyond Neptune, of which Pluto would be just the first. The next one was not found until 1992 but, since then, hundreds have been identified, including Eris in 2005. Some of the smaller ones transform into comets (p.58) when they stray nearer the Sun. This artist's impression is based on what is known about Eris, which has a small moon called Dysnomia.

THE DISCOVERY OF TRITON
The moon Triton was discovered in 1846. It interests scientists for several reasons. It has a retrograde orbit around Neptune – that is, the moon moves in the opposite direction in which the planet rotates. It is also the coldest object in the Solar System with a temperature of − 235°C (–391°F). Despite its low temperatures, Triton is a fascinating world. It has a pinkish surface, probably made of methane ice, which has repeatedly melted and refrozen. It has active volcanoes that spew nitrogen gas and darkened methane ice high into the thin atmosphere.

Sea-blue atmosphere

Ocean of water and gas

Smaller dark spot

PLUTO AND CHARON
The distance between Pluto and its moon, Charon, is only 19,700 km (12,240 miles). Charon was discovered in 1978 by a study of images of Pluto that looked suspiciously elongated. The clear image on the right was taken by the Hubble telescope (p.7), which allows better resolution than anything photographed from the Earth (left).

CLOSE-UP OF NEPTUNE
This picture was taken by *Voyager* 2 in 1989 after its 12-year voyage through the Solar System. It was 6 million km (3.8 million miles) away. *Voyager* went on to photograph the largest moon, Triton, and to reveal a further six moons orbiting the planet. Neptune has a beautiful, sea-blue atmosphere, composed mainly of hydrogen and a little helium and methane. This covers a huge internal ocean of warm water and gases – appropriate for a planet named after the god of the sea. (Many French astronomers had wanted the new planet to be named "Le Verrier", in honour of its discoverer.) *Voyager* 2 discovered several storm systems on Neptune, as well as beautiful white clouds high in the atmosphere.

FACTS ABOUT PLUTO

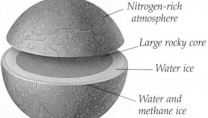

Nitrogen-rich atmosphere

Large rocky core

Water ice

Water and methane ice

- **Sidereal period** 248 Earth years
- **Temperature** −225°C
- **Rotational period** 6 days 9 hr
- **Mean distance from the Sun** 5,870 million km/ 3,650 million miles
- **Volume** (Earth = 1) 0.006
- **Mass** (Earth = 1) .0022 • **Density** (Water = 1) 2.03
- **Equatorial diameter** 2,390 km/1,485 miles
- **Number of satellites** 3

Travellers in space

NOT ALL MATTER IN THE SOLAR SYSTEM has been brought together to form the Sun and the planets. Clumps of rock and ice travel through space, often in highly elliptical orbits that carry them towards the Sun from the far reaches of the Solar System. Comets are icy planetary bodies that take their name from the Greek description of them as *aster kometes*, or "long-haired stars". Asteroids are mainly bits of rock that have never managed to come together as planets. However, Ceres, the largest by far at 940 km (585 miles) across, is like a little planet and since 2006 has been classed as a dwarf planet. A meteor is a piece of space rock – usually a small piece of a comet – that enters the Earth's atmosphere. As it falls, it begins to burn up and produces spectacular fireworks. A meteor that survives long enough to hit the Earth – usually a stray fragment from the asteroid belt – is called a meteorite.

PREDICTING COMETS
Going through astronomical records in 1705, Edmond Halley (1656–1743) noticed that three similar descriptions of a comet had been recorded at intervals of 76 years. Halley used Newton's recently developed theories of gravity and planetary motion (p.21) to deduce that these three comets might be the same one returning to the Earth at regular intervals, because it was travelling through the Solar System in an elliptical orbit (p.13). He predicted that the comet would appear again in 1758, but he did not live to see the return of the comet that bears his name.

Comet's orbit

Sun

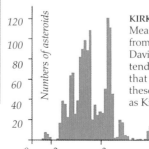

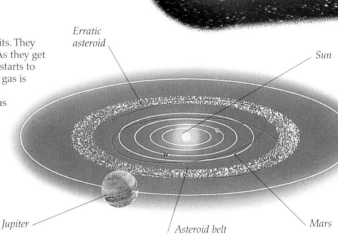

Nucleus

COMET'S TAIL
Comets generally have elongated orbits. They can be seen by the light they reflect. As they get closer to the Sun's heat, their surface starts to evaporate and a huge tail of dust and gas is given off. This tail always points away from the Sun because the dust and gas particles are pushed by solar wind and radiation pressure.

Comet's tail

Erratic asteroid

Sun

Jupiter

Asteroid belt

Mars

POSITION OF THE ASTEROID BELT
Since the Sicilian monk Guiseppe Piazzi discovered the first asteroid in January 1801, nearly 200,000 asteroids have been confirmed and numbered. Most of them travel in a belt between Mars and Jupiter, but Jupiter's great gravitational influence has caused some asteroids to swing out into erratic orbits.

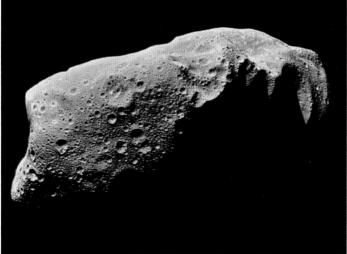

CLOSE-UP OF AN ASTEROID
This photograph of the asteroid Ida was taken by the *Galileo* spacecraft in 1993 as the space probe travelled to Jupiter. The cratered surface probably resulted from collisions with smaller asteroids. Ida is 52 km (32 miles) long.

KIRKWOOD GAPS
Measuring the distances of the known asteroids from the Sun in 1866, the American astronomer David Kirkwood (1814–1895) noticed that they tended to travel in loosely formed bands and that there were large, peculiar gaps between these bands. The gaps, which are now known as Kirkwood gaps, are due to recurring "bumps" from Jupiter's gravitational field. Asteroids can be catapulted into the inner Solar System by Jupiter's gravity.

Numbers of asteroids

120
100
80
60
40
20
0

2   3   4   5

Distance from Sun in astronomical units

HALLEY'S COMET FROM GIOTTO
When Halley's comet returned in 1986, the space probe *Giotto* was sent out to intercept and study it. The probe flew within 960 km (600 miles) of the comet, took samples of the vapour in its tail, and discovered that its nucleus was a jagged lump of dirt and ice measuring 16 x 8 km (10 x 5 miles).

Dust tail

Plasma (gas) tail

Meteorites

It was not until 1803 that the scientific community accepted that meteorites did, indeed, fall from space. Over 95 per cent of all the meteorites recovered are stone meteorites. Meteorites are divided into three types with names that describe the mix of elements found within each specimen. Stony meteorites look like stones but usually have a fused crust caused by intense heating as the meteorite passes through the Earth's atmosphere. Iron meteorites contain nickel-iron crystals, and stony iron meteorites are part stone, part iron.

MOLTEN DROPLET
Tektites are small, round, glassy objects that are usually the size of marbles. They are most often found on the Earth in great numbers, all together. When a blazing meteorite hits a sandstone region, the heat temporarily melts some of the minerals in the Earth's soil. These molten droplets harden to form tektites.

Tektite

METEORITE IN AUSTRALIA
This meteorite (left) fell near the Murchison River, Western Australia, in 1969. It contains significant amounts of carbon and water. The carbon comes from chemical reactions and not from once-living organisms like those carbon compounds found on Earth, such as coal.

Murchison meteorite

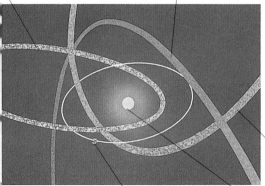

Geminids are seen in mid-December   *Perseids occur in mid-August*

Quadrantids are seen in early January

Earth   *Sun*

METEOR SHOWERS
When the Earth's orbit cuts through a stream of meteors, the meteoritic material seems to radiate out from one point in the sky, creating a meteor shower. The showers are given names, such as "the Geminids", that derive from the con-stellations in the sky from which they seem to be coming.

ICY CRATER
The Earth bears many scars from large meteorites, but the effects of erosion and vegetation cover over some of the spectacular craters. This space view shows an ice-covered crater near Quebec in Canada. It is now a 66-km (41-mile) wide reservoir used for hydroelectric power.

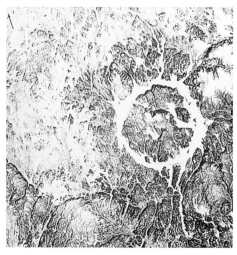

The birth and death of stars

APART FROM THE SUN, the closest star to the Earth is Proxima Centauri which is 4.2 light-years or 40 million, million km (25 million, million miles) away. A light-year is the distance that light or other electromagnetic radiation (p.32) travels in a year. Stars are luminous, gaseous bodies that generate energy by means of nuclear fusion in their cores (pp.38–39). As a star ages, it uses up its fuel. The core shrinks under its own weight while the nuclear "burning" continues. The shrinkage heats up the core, making the outer layers of the star expand and cool. The star becomes a "red giant". As the remains of the star's atmosphere escape, they leave the core exposed as a "white dwarf". The more massive stars will continue to fuse all their lighter elements until they reach iron. When a star tries to fuse iron, there is a massive explosion and the star becomes a "supernova". After the explosion, the star's core may survive as a pulsar or a "black hole" (p.62).

CATALOGUE OF NEBULAE
The French astronomer Charles Messier (1730–1817) produced a catalogue of around 100 fuzzy or nebulous objects in 1784. Each object was numbered and given an "M-" prefix. For example, the Orion nebula, the 42nd object in Messier's list, is referred to as M42. Many of the objects he viewed were actually galaxies and star clusters.

HENRIETTA LEAVITT (1868–1921)
In 1912 the American astronomer Henrietta Leavitt was studying Cepheid variable stars. These are a large group of bright yellow giant and supergiant stars named after their prototype in the constellation of Cepheus. Variable stars are stars that do not have fixed brightness. Leavitt discovered that the brighter stars had longer periods of light variation. This variation can be used to determine stellar distances beyond 100 light-years.

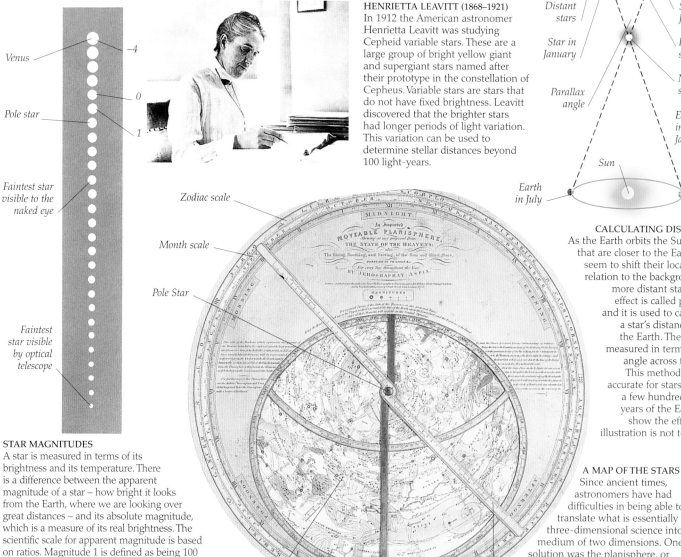

Venus

−4

0

1

Pole star

Faintest star visible to the naked eye

Faintest star visible by optical telescope

Zodiac scale

Month scale

Pole Star

Area of sky visible to viewer

Rule

Distant stars

Star in July

Star in January

Parallax shift

Nearby star

Parallax angle

Earth in January

Sun

Earth in July

STAR MAGNITUDES
A star is measured in terms of its brightness and its temperature. There is a difference between the apparent magnitude of a star – how bright it looks from the Earth, where we are looking over great distances – and its absolute magnitude, which is a measure of its real brightness. The scientific scale for apparent magnitude is based on ratios. Magnitude 1 is defined as being 100 times brighter than Magnitude 5. In this scale, the punched holes show the brightest star at the top and the faintest at the bottom.

CALCULATING DISTANCE
As the Earth orbits the Sun, stars that are closer to the Earth will seem to shift their location in relation to the background of more distant stars. This effect is called parallax and it is used to calculate a star's distance from the Earth. The shift is measured in terms of an angle across the sky. This method is only accurate for stars within a few hundred light-years of the Earth. To show the effect the illustration is not to scale.

A MAP OF THE STARS
Since ancient times, astronomers have had difficulties in being able to translate what is essentially a three-dimensional science into the medium of two dimensions. One solution was the planisphere, or "flattened sphere", in which the whole of the heavens was flattened out with the Pole Star at the centre of the chart.

STUDYING THE STARS

The British astronomer Williams Huggins (1824–1910) was one of the first to use spectroscopy for astronomical purposes (pp.30–31). He was also the first astronomer to connect the Doppler effect (which relates to how sound travels) with stellar red shift (p.23). In 1868 he noticed that the spectrum of the bright star Sirius has a slight shift towards the red end of the spectrum. Although his measurement proved spurious, he correctly deduced that this effect is due to that star travelling away from the Earth.

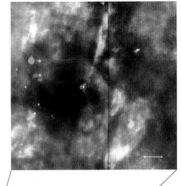

NOVAE AND SUPERNOVAE

Novae and supernovae are stars that suddenly become much brighter, then gradually fade. Novae are close double stars in which material dumped onto a white dwarf from its partner detonates a nuclear explosion. Supernovae are even brighter and more violent explosions. One type is triggered like a nova but the nuclear explosion destroys the white dwarf. A supernova also occurs when the core of a massive dying star collapses. The core may survive as a neutron star or black hole. The gas blown off forms an expanding shell called a supernova remnant. These pictures show supernova 1987A before (right) and after (left) it exploded.

BETELGEUSE

Betelgeuse is a variable star that is 17,000 times brighter than our Sun. It lies on the shoulder of Orion the Hunter, 400 light-years from the Earth. Astronomers believe that it will "die" in a supernova explosion (above right).

Outline of Orion, the Hunter

Bellatrix

THE STELLAR NURSERY

The material in a nebula – a stellar nursery made up of gases and dust – collapses under gravity and eventually creates a cluster of young stars. Each star develops a powerful wind, which clears the area to reveal the star surrounded by a swirling disc of dust and gas. This may form a system of planets or blow away into space.

Rigel

THE CONSTELLATION ORION

A constellation is a group of stars that appear to be close to each other in the sky, but that are usually spread out in three-dimensional space. Orion the Hunter's stars include the bright Betelgeuse and Rigel.

ORION NEBULA M42

Stars have a definite life cycle that begins in a mass of gas that turns into stars. This "nebula" glows with colour because of the cluster of hot, young stars within it. This is part of the Great Nebula in Orion.

Our galaxy and beyond

THE FIRST STARS WERE FORMED a few hundred million years after the Universe was born. Clumps containing a few million brilliant young stars merged to form galaxies. A typical galaxy contains about 100 billion stars and is around 100,000 light-years in diameter. Edwin Hubble was the first astronomer to study these distant star systems systematically. While observing the Andromeda Galaxy in 1923, he was able to measure the brightness of some of the stars in it, although his first estimate of their distance was incorrect. After studying the different red shifts of the galaxies (p.23), Hubble proposed that the galaxies are moving away from our galaxy at speeds proportional to their distances from us. His law shows that the Universe is expanding.

EDWIN HUBBLE (1889–1953)
In 1923 the American astronomer Edwin Hubble studied the outer regions of what appeared to be a nebula (p.61) in the constellation of Andromeda. With the high-powered 254-cm (100-in) telescope at Mount Wilson, he was able to see that the "nebulous" part of the body was composed of stars, some of which were bright, variable stars called Cepheids (pp.60–61). Hubble realized that for these intrinsically bright stars to appear so dim, they must be extremely far away from the Earth. His research helped astronomers to begin to understand the immense size of the Universe.

THE MILKY WAY
From the Earth, the Milky Way appears particularly dense in the constellation of Sagittarius because this is the direction of the galaxy's centre. Although optical telescopes cannot penetrate the galactic centre because there is too much interstellar dust in the way, radio and infrared telescopes can.

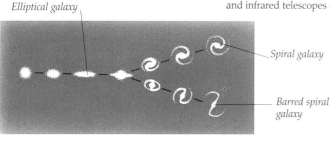

Elliptical galaxy

Spiral galaxy

Barred spiral galaxy

CLASSIFYING GALAXIES
Hubble devised a classification of galaxies according to shape. Elliptical galaxies were subdivided by how flat they appeared. He classified spiral and barred spiral galaxies (where the arms spring from a central bar) according to the tightness of their arms.

The Milky Way photographed from Chile with a wide-angle lens

Observatory building

WHIRLPOOL GALAXY
The Whirlpool Galaxy is a typical spiral galaxy, approximately 25 million light-years away. It can be found in the faint constellation Canes Venatici, at the end of the tail of the constellation of Ursa Major, or the Great Bear. It was one of the nebulae drawn by the third Earl of Rosse (p.26) in the 19th century.

THE SHAPE OF OUR GALAXY

The Milky Way, seen edge-on (top), has an oval central bulge surrounded by a very thin disc containing the spiral arms. It is approximately 100,000 light-years in diameter and about 15,000 light-years thick at its centre. Our Sun is located about 30,000 light-years away from the centre. The Milky Way looks like a band in our skies because we see it from "inside"– its disc is all around us. Viewed from above (bottom), it is a typical spiral galaxy with the Sun situated on one of the arms, known as the Orion arm.

Sun

Central plane

Sun on Orion's arm

Central plane

Horizon

ANDROMEDA GALAXY

The Andromeda Galaxy is a spiral galaxy, shaped like our Milky Way, but it has nearly half as much mass again. It is the most distant object that is visible to the unaided eye. It has two small elliptical companion galaxies.

What is cosmology?

Cosmology is the name given to the branch of astronomy that studies the origin and evolution of the Universe. It is an ancient study, but in the 20th century the theory of relativity, advances in particle physics and theoretical physics, and the discoveries about the expanding Universe gave cosmology a more scientific basis and approach.

ALBERT EINSTEIN (1879–1955)

In proposing that mass is a form of energy, the great German-American scientist Albert Einstein redefined the laws of physics dominant since Newton's time (p.21). The fact that gravitation could affect the shape of space and the passage of time meant that scientists were finally provided with the tools to understand the birth and death of the stars, especially the phenomenon of the black hole.

Massive star

Black hole　*X-rays*

BLACK HOLES

A supernova (p.61) can leave behind a black hole – an object that is so dense and so collapsed that even light cannot escape from it. Although black holes can be detected when gas spirals into them, because the gases emit massive quantities of X-rays as they are heated, they are otherwise very hard to find. Sometimes they act as "gravitational lenses", distorting background starlight.

Did you know?

AMAZING FACTS

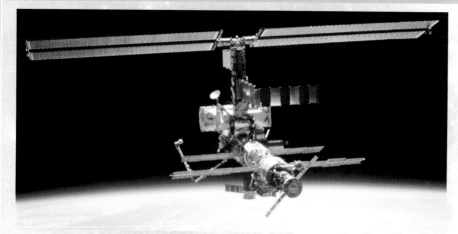

The International Space Station

Designed to carry out invaluable research, work on the International Space Station (ISS) started in 1998. Due for completion in 2010, it is being built entirely in orbit, involving spacewalks by astronauts and the use of space robotics.

Since 1995, astronomers have discovered hundreds of planetary systems around ordinary stars. The star 55 Cancri, which is similar to the Sun and 41 light-years away, has a family of at least five planets similar to the giant planets of the solar system.

Some galaxies are cannibals – they consume other galaxies! Hubble has taken pictures of the Centaurus A galaxy. At its centre is a black hole that is feeding on a neighbouring galaxy.

Jupiter's moon, Europa, has an ocean of liquid water or slush under its icy crust. Parts of the surface look as if great rafts of ice have broken up and moved about.

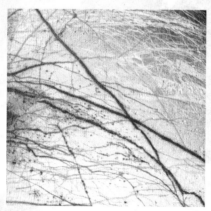

A colour-enhanced view of Europa's surface

Wolf-Rayet stars are among the hottest, most massive stars known and one of the rarest types. At least 25 times bigger than the Sun, and with temperatures up to 100,000 °C (180,000 °F), they are close to exploding as phenomenally powerful supernovae.

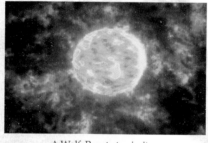

A Wolf-Rayet star in its final hours

The world's largest optical telescopes are the Keck Telescopes in Hawaii. Each one is the height of an eight-storey house.

The twin Keck Telescopes on Mauna Kea, Hawaii

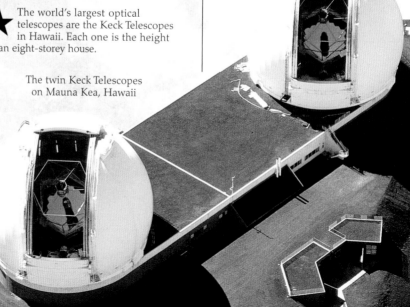

The *Pioneer 10* spacecraft, which was launched in 1972, is still transmitting signals back to Earth, although NASA stopped monitoring them in 1997. Now more than 11 billion km (7 billion miles) away, *Pioneer 10* should reach the stars in the constellation of Taurus in about two million years.

Apollo (USA) and *Lunas 17* and *21* (Russia) have brought back samples of rocks from the surface of the Moon. Some of these rocks are up to 4.5 billion years old – older than any rocks found on Earth.

The Large Magellanic Cloud is a galaxy that orbits the Milky Way. It contains a dazzling star cluster known as NGC 1818, which contains over 20,000 stars, some of which are only about a million years old.

In 2000, scientists identified the longest comet tail ever. Comet Hyakutake's core was about 8 km (5 miles) across – but its tail measured over 570 million km (350 million miles) long.

When Pluto was discovered in 1930, it was given its name as a result of a suggestion made by 11-year-old English schoolgirl Venetia Burney.

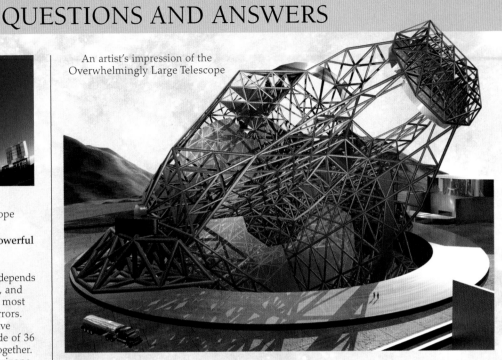

An artist's impression of the Overwhelmingly Large Telescope

The four telescopes that make up the Very Large Telescope

Q Which are the world's most powerful telescopes?

A The performance of a telescope depends on the total area of its mirrors, and its ability to distinguish detail. The most powerful combine two or more mirrors. The twin Keck Telescopes (p.64) have mirrors 10 m (33 ft) across, each made of 36 hexagonal segments, and can work together. The Large Binocular Telescope in Arizona has two 8.4-m (27.6-ft) mirrors. Together they work like an 11.8-m (38.7-ft) mirror. The Very Large Telescope in Chile (above) consists of four 8.2-m (27-ft) telescopes that can observe together or separately.

Q How far can astronomers see?

A The most distant galaxies so far detected are about 13 billion light-years away. This means that they were formed only a few hundred million years after the Big Bang.

Q How big will telescopes be in the future?

A The Overwhelmingly Large Telescope (Owl), may be built in the Atacama Desert, Chile, and be operational around 2017. Its main mirror, made up of hexagonal segments, would be over 100 m (330 ft) across. Owl's designers hope to make use of both active and adaptive optics to achieve the best possible resolution. Active optics adjust the mirror segments so they work as a single sheet of glass. Adaptive optics work by shining a powerful laser into the sky, then adjusting the mirrors to keep the laser sharply in focus. This lets the telescope correct for distortions caused by the atmosphere.

Q Is Hubble the only space telescope?

A Launched in 1990, the Hubble Space Telescope (HST) was the first major observatory in space and is the most famous, but many others have operated in orbit around the Earth or Sun. They are usually designed to last for several years and to make particular kinds of observations. Some of the most important are NASA's four "Great Observatories", of which HST is one. The others are the Compton Gamma-ray Observatory, which operated between 1991 and 2000, the Chandra X-ray Observatory, launched in 1999, and the Spitzer Space Telescope. Spitzer is an infrared telescope launched in 2003. The successor to HST will be the James Webb Space Telescope, due for launch in about 2013.

Q How does space technology help us find our way on Earth?

A As well as helping us map the Universe, satellites are also improving our ability to navigate on Earth. The Global Positioning System (GPS) is a collection of 27 satellites that are orbiting the Earth – 24 in action and three back-ups. Their orbits have been worked out so that at any time there are at least four of them visible from any point on the planet. The satellites constantly broadcast signals that indicate their position. These signals can be picked up by devices called GPS receivers. A receiver compares the information from the satellites in its line of sight. From this, it can work out its own latitude, longitude, and altitude – and so pinpoint its position on the globe. GPS technology has some amazing applications. It is already being used in cars. By linking a GPS receiver to a computer that stores data such as street maps, an in-car system can plot the best route to a particular location.

GPS car navigation system

Record Breakers

☾ **USING A TELESCOPE**
The first astronomers to study the night sky through a telescope were Thomas Harriott (1560–1621) and Galileo Galilei (1564–1642).

☾ **LARGEST INFRARED TELESCOPE**
The mirror of the Hobby-Eberly Telescope (HET) on Mount Fowlkes in Texas, USA, is 11 m (36 ft) across.

☾ **PLANET WITH THE MOST MOONS**
Jupiter's moons numbered at least 63 at the last count – but astronomers are still finding new ones.

☾ **HIGHEST VOLCANO**
At 26,400 m (86,600 ft) high, Olympus Mons on Mars is the highest volcano in our Solar System.

☾ **NEAREST NEBULA**
The nearest nebula to Earth is the Merope Nebula, 380 light-years away.

Cutting-edge astronomy

FINDING OUT HOW THE UNIVERSE was born and has evolved is a great challenge for astronomers. In 1964 the Universe was found to be full of radiation, predominantly microwaves. This is called the cosmic microwave background (CMB) and is a relic from the "Big Bang" when the Universe began. It gives a glimpse of the Universe as it was when just a few hundred thousand years old. Since the Big Bang, the Universe has been expanding and this expansion has been speeding up for the last five billion years, driven by a mysterious force – "dark energy".

Back-to-back dishes scan deep space

Solar array shields telescope from Sun's heat

Instrument cylinder houses the equipment

SOUTH POLE INTERFEROMETER
The Degree Angular Scale Interferometer (DASI) spent several years measuring in fine detail how the cosmic microwave background radiation varies across the sky. It was sited at the Amundsen-Scott research station at the South Pole. The freezing temperatures there keep the atmosphere nearly free of water vapour, which is important for detecting microwaves.

MICROWAVE ANISOTROPY PROBE
The Wilkinson Microwave Anisotropy Probe (WMAP) was launched by NASA in 2001 on a mission of about six years to survey the cosmic microwave background radiation with unprecedented accuracy. It was put in an orbit around the Sun, on the opposite side of the Earth from the Sun, and four times farther away than the Moon. From this vantage point it could view the whole sky without interference.

BOOMERANG TELESCOPE
In 1998, a microwave telescope nicknamed Boomerang flew over Antarctica for ten days at an altitude of 37,000 m (121,000 ft). The picture it sent back of around three per cent of the sky showed regular patterns in the CMB. They appear to be shock waves travelling through the young Universe – perhaps even echoes of the Big Bang.

Giant weather balloon filled with helium to float at very high altitudes

The telescope waits to be lifted into the upper atmosphere

QUASAR HOME GALAXY

Quasars are so bright, the galaxies surrounding them are overwhelmed by their light and are almost invisible in normal images. The Hubble Space Telescope used a specially equipped camera to block the glare from the quasar 3C 273 and make this image (left) of the much fainter galaxy surrounding it.

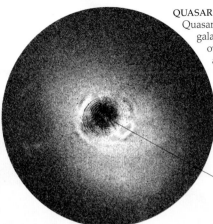

Black disk in camera blocks glare from quasar

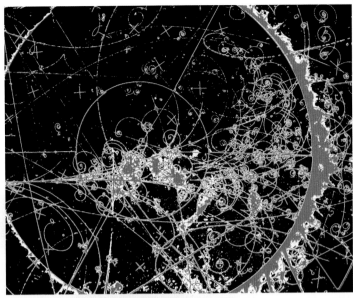

NEUTRINOS

This bubble chamber shows the pattern left by a sub-atomic particle called a neutrino after a high-speed collision. Huge numbers of neutrinos reach Earth from space but they are very difficult to detect. Studying them helps to understand nuclear processes in stars.

BIG BANG

BIG BANG

The CMB map is a strong piece of evidence for the Big Bang theory. If the Universe was once extremely hot and dense, and then cooled and expanded, a point should be reached when radiation would separate from matter. The trace this would leave behind matches the CMB.

CMB MAP

This map of the sky is laid out like any map of a globe – it shows the view in all directions from Earth. The most exciting thing for astronomers is that it is not perfectly uniform, but has patterns that give us clues about the forces at work in the Universe around 13,000 million years ago.

Dark areas are cooler by a tiny percentage

Hot spots show where galaxies will form

Cosmic Microwave Background as seen by the WMAP probe

STEPHEN HAWKING

Cosmologist Stephen Hawking worked with Roger Penrose to show how Einstein's theories about space and time support the Big Bang idea. Hawking used similar calculations to predict that black holes should be found to emit radiation.

Find out more

IF YOU WANT TO KNOW MORE about astronomy, just look at the sky! Binoculars or telescopes help, but there are around 2,500 stars that are visible to the naked eye. Invest in a pocket-sized guide to the constellations, so that you can identify what you observe. You can find tips on what to look for on a particular night at astronomy websites, on specialist television programmes, and even in some newspapers. You can also fuel your star-gazing hobby by visiting science museums, which have lots of displays on space science of the past and the future.

ASTRONOMY ON TV
Television programmes are a good introduction to the night sky. *The Sky at Night* is the world's longest-running astronomy programme. In 1959, its presenter, Patrick Moore, showed audiences the first pictures of the far side of the Moon.

TV astronomer Patrick Moore

Without tent light, stars would appear even brighter and clearer

VISITING A PLANETARIUM
At a planetarium, stunning footage of the cosmos from world-class telescopes is projected on to a domed screen above your head. The planetarium shown below is in Brittany, France.

VIEWING THE NIGHT SKY
Anyone can be a stargazer. Using a standard pair of binoculars, this amateur astronomer in the USA has a great view of the Milky Way. The stars above the tent are in the constellation Sagittarius.

A seamless screen makes viewers feel like they are really viewing the sky

ALIEN LIFE?
Perhaps one day astronomers will find definite proof of alien life. In the meantime, there are plenty of movies about other life forms. Of course they are pure fiction, and "real" aliens would probably look nothing like the movie versions, but they are still great fun to watch!

The alien stars of *Mars Attacks!* (1996)

WHITE NOISE
You can see traces of the Cosmic Microwave Background just by turning on your television! When it is tuned between channels, the "snow" you see is partly microwave radiation from space.

WAITING FOR AN ECLIPSE

Eclipses are amazing events, but be sure to protect your eyes from the Sun's dangerous rays if you are lucky enough to see one. In any year, there can be up to seven solar or lunar eclipses. Although solar eclipses are more common, they seem rarer because they are only ever visible in a narrow area.

THE GIBBOUS MOON

The Moon is the ideal starting point for the amateur astronomer. Over a month you can observe each of its phases. This Moon, superimposed onto a photograph of Vancouver, Canada, is gibbous – that is, more than half full.

Places to visit

ADLER PLANETARIUM AND ASTRONOMY MUSEUM, CHICAGO, USA
- Virtual-reality experiences of the Universe
- Historical astronomical instruments

ARMAGH PLANETARIUM, NORTHERN IRELAND, UK
- Astronomical shows to suit all ages

JODRELL BANK SCIENCE CENTRE, MACCLESFIELD, CHESHIRE, UK
- Home of the Lovell Telescope
- Hands-on exhibits
- 150-seater planetarium

NATIONAL MARITIME MUSEUM AND ROYAL OBSERVATORY, GREENWICH, UK
- Over two million objects, including a vast collection of astronomical instruments

NATIONAL SPACE SCIENCE CENTRE, LEICESTER, UK
- Five themed galleries of interactive exhibits
- Planetarium

PARC DE LA VILLETTE, PARIS, FRANCE
- Interactive exhibits plus an *Ariane* rocket
- Planetarium

SCIENCE MUSEUM, LONDON, UK
- Over 300,000 objects, including the *Apollo 10* Command Module

Bronze equatorial sundial

ADLER PLANETARIUM AND ASTRONOMY MUSEUM

Most science museums have galleries dedicated to astronomy, but there are also specialist museums. Chicago's Adler Astronomy Museum has multimedia exhibits and an impressive array of antique telescopes.

USEFUL WEBSITES

- An excellent introduction from NASA for amateur astronomers
  http://spacekids.hq.nasa.gov/
- Up-to-the-minute tips on what to look for in the night sky from the expert astronomers at Jodrell Bank
  www.jb.man.ac.uk/public/nightsky.html
- The website of the UK's largest amateur astronomy society
  www.popastro.com
- A different astronomical picture to look at every day
  http://antwrp.gsfc.nasa.gov/apod/astropix.html

Glossary

Aurora borealis

APHELION The point in a planet's orbit where it is farthest from the Sun.

ASTEROID A chunk of planet material in the Solar System.

ASTROLOGY The prediction of human characteristics or activities according to the motions of the stars and planets.

ASTRONOMICAL UNIT (AU) The average distance between the Earth and the Sun – 150 million km (93 million miles).

ASTRONOMY The scientific study of the stars, planets, and Universe as a whole.

ASTROPHYSICIST Someone who studies the way stars work.

ATMOSPHERE The layer of gases held around a planet by its gravity. Earth's atmosphere stretches 1,000 km (600 miles) into space.

ATOM Smallest part of an element, made up of subatomic particles, such as protons, electrons, and neutrons.

Charge coupled device (CCD)

AURORA Colourful glow seen in the sky near the poles, when electrically-charged particles hit gases in the atmosphere.

AXIS Imaginary line through the centre of a planet or star, around which it rotates.

BIG BANG Huge explosion that created the Universe around 13,000 million years ago.

BLACK HOLE A collapsed object with such powerful gravity that nothing can escape it.

CHARGE COUPLED DEVICE (CCD) Light-sensitive electronic device used for recording images in modern telescopes.

COMET An object of ice and rock. When it nears the Sun, it has a glowing head of gas with tails of dust and gas.

CONCAVE Curving inwards.

CONSTELLATION The pattern that a group of stars seems to make in the sky.

CONVEX Curving outwards.

CORONA The Sun's hot upper atmosphere.

CORONAGRAPH A telescope used to observe the edge (corona) of the Sun.

COSMIC BACKGROUND RADIATION (CBR) A faint radio signal left over from the Big Bang.

COSMOS The Universe.

DOPPLER EFFECT The change in a wave frequency when a source is moving towards or away from an observer.

ECLIPSE When one celestial body casts a shadow on another. In a lunar eclipse, Earth's shadow falls on the Moon. In a solar eclipse, the Moon casts a shadow on Earth.

ECLIPTIC Imaginary line around the sky along which the Sun appears to move.

ELECTROMAGNETIC RADIATION Waves of energy that travel through space at the speed of light.

ELECTROMAGNETIC SPECTRUM The complete range of electromagnetic radiation.

EQUINOX Twice-yearly occasion when day and night are of equal length, falling on about 21 March and 23 September.

FOCAL LENGTH The distance between a lens or mirror and the point where the light rays it collects are brought into focus.

FOSSIL The naturally preserved remains of animals or plants, or evidence of them.

FREQUENCY The number of waves of electromagnetic radiation that pass a point every second.

GALAXY A body made up of millions of stars, gas, and dust, held together by gravity.

A communications satellite in geostationary orbit

GAMMA RAY Electromagnetic radiation with a very short wavelength.

GEOLOGIST Someone who studies rocks.

GEOSTATIONARY ORBIT An orbit 35,880 km (22,295 miles) above the Equator, in which a satellite takes as long to orbit Earth as Earth takes to spin on its axis.

GRAVITY Force of attraction between any objects with mass, such as the pull between the Earth and the Moon.

INFRARED Type of electromagnetic radiation also known as heat radiation.

LATITUDE Position to the north or south of the Equator, in degrees.

LIBRATION A wobble in the Moon's rotation that allows observers to see slightly more than half its surface.

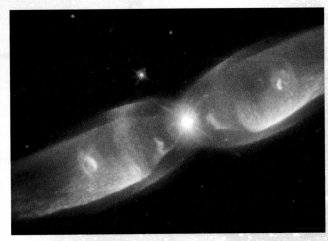
M2–9 planetary nebula

NEUTRON STAR A collapsed star left over after a supernova.

NOVA A white dwarf star that suddenly flares up and shines about 1,000 times brighter than before, after receiving material from a companion star.

NUCLEAR FUSION When the nuclei (centres) of atoms combine to create energy.

OBSERVATORY A place where astronomers study space.

SIDEREAL TIME Time measured by the stars rather than by the Sun.

SOLAR SYSTEM Everything held by the Sun's gravity, including planets and comets.

SOLSTICE Twice-yearly occasion when the Sun is farthest from the Equator, falling on about 21 June and 21 December,

SPECTROSCOPY The study of the spectrum of a body that emits radiation.

STAR A hot, massive, shining ball of gas that makes energy by nuclear fusion.

SUBATOMIC PARTICLE Particle smaller than an atom, for example a proton, neutron, or electron.

SUNSPOT A cool dark spot on the Sun's surface, created by the Sun's magnetic field.

SUPERNOVA An enormous explosion, created when a supergiant star runs out of fuel, or when a white dwarf explodes.

TIDE The regular rise and fall of the sea caused by the gravitational pull of the Sun and the Moon on the Earth.

LIGHT-YEAR The distance light travels in a year – around 9.5 million million km (5.9 million million miles).

LONGITUDE Position to the east or west of the Greenwich Meridian, in degrees.

MASS A measure of the amount of matter in an object and how it is affected by gravity.

MATTER Anything that has mass and occupies space.

MERIDIAN An imaginary line linking the poles. The one at Greenwich marks 0 degrees.

METEOR The streak of light seen when comet dust burns up as it enters the Earth's atmosphere.

METEORITE A fragment of space rock that has fallen on to a planet or moon.

METEOROLOGICAL To do with weather.

MICROWAVE The type of radio wave that has the shortest radio wavelengths.

NEBULA A cloud of dust and gas in space.

NEUTRINO A subatomic particle produced by nuclear fusion in stars or the Big Bang.

OCCULTATION When one heavenly body passes before another, hiding it from view.

OORT CLOUD Huge spherical comet cloud, about 1.6 light-years wide, that surrounds the Sun and planets.

ORBIT The path of one object around another more massive object in space.

PARALLAX Shift in a nearby object's position against a more distant background when seen from two separate points, used to measure the distance of nearby stars.

PAYLOAD Cargo carried by a space vehicle or an artificial satellite.

PERIHELION The point in an object's orbit where it is closest to the Sun.

PHASE Size of the illuminated part of a planet or moon seen from Earth.

PHOTOSPHERE A star's visible surface, from which its light shines.

PLANET Large globe of rock, liquid, or gas that orbits a star.

PRISM A transparent block used to change the direction of a beam of light.

PROMINENCE A huge arc of gas in the Sun's lower corona.

PULSAR A spinning neutron star.

QUASAR A distant active galaxy releasing lots of energy from a central small area.

RADIO TELESCOPE Telescope that detects radio waves from objects in space.

REFLECTING TELESCOPE Telescope that gathers light with a concave mirror.

REFRACTING TELESCOPE Telescope that gathers light with a combination of lenses.

SATELLITE Any object held in orbit around another object by its gravity, including moons and artificial satellites.

The Sun, through a filter, shows prominences as dark streaks

ULTRAVIOLET Electromagnetic radiation with a shorter wavelength than visible light.

VACUUM A perfectly empty – or very nearly empty – space.

WAVELENGTH Distance between the peaks or troughs in waves of radiation.

X-RAY Electromagnetic radiation with a very short wavelength.

ZODIAC The 12 constellations through which the Sun, Moon, and planets appear to move.

...ar with magnetic field shown in purple

Index

Acknowledgements

Dorling Kindersley would like to thank:
Maria Blyzinsky for her invaluable assistance in helping with the objects at the Royal Observatory, Greenwich; Peter Robinson & Artemi Kyriacou for modelling; Peter Griffiths for making the models; Jack Challoner for advice; Frances Halpin for assistance with the laboratory experiments; Paul Lamb, Helen Diplock, & Neville Graham for helping with the design of the book; Anthony Wilson for reading the text; Harris City technology College & The Royal Russell School for the loan of laboratory equipment; the Colour Company & the Roger Morris Partnership for retouching work; lenses supplied by Carl Lingard Telescopes, 89 Falcon Crescent, Clifton, Swinton, Manchester; Jane Parker for the index; Stewart J. Wild for proof-reading; David Ekholm–JAlbum, Sunita Gahir, Susan Reuben, Susan St. Louis, Lisa Stock, & Bulent Yusuf for the clipart; Neville Graham, Sue Nicholson, & Susan St. Louis for the wallchart.

Illustrations Janos Marffy, Nick Hall, John Woodcock and Eugene Fleury

Photography Colin Keates, Harry Taylor, Christi Graham, Chas Howson, James Stevenson and Dave King.

Picture credits
t=top b=bottom c=centre l=left r=right

American Institute of Physics: Emilio Segrè Visual Archives/Bell Telephone Laboratories 32cl; Research Corporation 32bl; Shapley Collection 60cl; Ancient Art and Architecture Collection: 9tl, 9cr, 20tl; Anglo-Australian Telescope Board: D. Malin 61tr; Archive für Kunst and Geschichte, Berlin: 19tl; National Maritime Museum 46c; Associated Press: 8tl; The Bridgeman Art Library: 28tl Lambeth Palace Library, London 17tr; The Observatories of the Carnegie Institution of Washington: 39cr; Jean-Loup Charmet: 60tl; Bruce Coleman Ltd: 43c; Corbis: Russell Christophe/Kipa 69tl; Sandy Felsenthal 69r; Hulton-Deitsch Collection 68tl; ESA: 48br; CNES/Arianespace 35tl; ET Archive: 9tr; European Southern Observatory: 65tr; Mary Evans Picture Library: 6tl, 14bl, 18cb, 42tl, 61tcl; Galaxy Picture Library: Boomerang Team 66b; JPL 64bl; MSSS 49tr, 49br; Margaret Penston 67br; Robin Scagell 68bc; STScI 67tl; University of Chicago 66cl; Richard Wainscoat 27tl, 64br; Gemini Observatory: Neelon Crawford/Polar Fine Arts/US National Science Foundation 25cl; Ronald Grant Archive: 68cl; Robert Harding: R. Frerck 32bc; C. Rennie 10br; Hulton Deutsch: 20c; Henry E. Huntington Library and Art Gallery 62tl; Images Colour Library 7tl, 7cr, 7c, 16tl, 16cl, 18cl; Image Select 19br, 21tc, 23tl, 26cl, 28cr; JPL 3cr, 37cr, 49c, 49cr, 50tr, 51crb, 52cl, 52br, 55tr, 56cl, 63tr; Lowell Observatory 48tr; Magnum: E. Lessing 19tr;

NASA: 3cr, 8cl, 35cr, 35cl, 35bl, 41cr, 44bl, 44cl, 46br, 47b, 56–57bc, 57tl; Dana Berry/Sky Works Digital 64c; ESA and Erich Karkoschka (University of Arizona) 53t; Hubble Space Telescope Comet Team 51t; JPL/Space Science Institute 34bc, 39tc, 49tl, 50br, 51bc, 53bc, 53c, 55c, 55cr, 56cb, 61cr; LMSAL 47tr; WMAP Science Team 66tr; National Geophysical Data Centre: NOAA 27cl; National Maritime Museum Picture Library: 6cl, l0cr, 15tl, 15tr, 25tl, 27br, 29tr, 54tl; National Radio Astronomy Observatory: AUI/ J M Uson 32tl; Novosti (London): 28br, 35cl, 47tr; Planétarium de Brétagne: 68-69; Popperfoto: 47tl; 67crb; Rex Features Ltd: 34clb; Scala/ Biblioteca Nationale: 20cr; Science Photo Library: 18tl, 25bc, 31tl, 31bl, 48c, 53bc, 56tl, 57cr; Dr. J. Burgess 26tl; Chris Butler 70-71 bckgrd, 71br; Cern 67tr; Jean-Loup Charmet 11tl, 37tl;J-C Cuillandre/Canada-France-Hawaii telescope 68-69 bckgrd; F. Espenak 46tr; European Space Agency 35tl, 42-43bc; Mark Garlick 57cra; GE Astrospace 70cr; Jodrell Bank 33tr; Mehau Kulyk 67c; Dr. M.J. Ledlow 33tl; Chris Madeley 70tl; F. D. Miller 31c; NASA 7crb, 20bc, 20br, 20br, 34cr, 41crb, 44-45bc, 49tc, 57cr, 58bl, 59br, 64t, 66-67 bckgrd, 67bc; NOAO 61cl, 71br; NASA/ESA/STScI/E. KARKOSCHKA, U.ARIZONA 54–55b; Novosti 47cr; David Nunuk 25br, 65tl, 69tc; David Parker 70bl; Physics Dept. Imperial College 30crb, 30br; P. Plailly 38cl; Philippe Psaila 65br; Dr. M. Read 30tl; Royal Observatory, Edinburgh 59tr; J. Sandford 41t, 61bl; Space Telescope

Science Institute/NASA 64-65 bckgrd, 71bl; Starlight/R. Ressemeyer 12tl, 27tr, 33cr, 55tl, 62-63cb; U.S. Geological Survey 37br, 481; Frank Zullo 68tr; SOHO (ESA & NASA): 38–39b; Tony Stone Images: 42cl; Roger Viollet/ Boyes 38tl; Zefa UK: 6-7bc, 8bl, 33b, 61br, 62bl; G. Heil 26bc.

With the exception of the items listed above, the object from the British Museum on page 8c, from the Science Museum on pages 21b, 22 cl, and from the Natural History Museum on page 43tl, the objects on pages 1, 2t, 2c, 2b, 3t, 31, 3b, 3tr, 4, 11b, 12b, 14c, 14bc, 14r, 15tl, 15b, 16bl, 16r, 17bl, 17br, 20bl, 24bc, 25tr, 28cl, 28c, 28bl, 29tl, 29c, 29b, 31b, 36b, 38b, 40cl, 40bl, 40/4b, 42bl, 52bl, 54bl, 58tr, 60b, are all in the collection of the Royal Observatory, Greenwich.

Wallchart credits: Corbis: Roger Ressmeyer crb (optical telescope), fcl; DK Images: ESA fc London Planetarium fcla, fcrb (Jupiter); Science Museum, London cla (Galileo's telescope); NASA: fbr, fcra (Moon); Science Photo Library: David Nunuk bl

All other images © Dorling Kindersley.
For further information, see:
www.dkimages.com